The Science Review Article

Linguistic Insights

Studies in Language and Communication

Edited by Maurizio Gotti,
University of Bergamo

Volume 17

PETER LANG
Bern · Berlin · Bruxelles · Frankfurt am Main · New York · Oxford · Wien

Judy Noguchi

The Science Review Article

An Opportune Genre in the Construction of Science

PETER LANG
Bern · Berlin · Bruxelles · Frankfurt am Main · New York · Oxford · Wien

Bibliographic information published by Die Deutsche Bibliothek
Die Deutsche Bibliothek lists this publication in the Deutsche Nationalbibliografie; detailed bibliographic data is available on the Internet at ‹http://dnb.ddb.de›.

British Library and Library of Congress Cataloguing-in-Publication Data:
A catalogue record for this book is available from *The British Library*, Great Britain, and from *The Library of Congress*, USA

ISSN 1424-8689
ISBN 3-03910-426-8
US-ISBN 0-8204-7034-1

Hochfeldstrasse 32, Postfach 746, CH-3000 Bern 9, Switzerland
info@peterlang.com, www.peterlang.com, www.peterlang.net

Contents

Preface 11

1. Discourse communities and scientific genres 19

1.1 "No language, no facts" 19
1.2 Different audiences, different sciences 22
1.3 From chaos, new gestalts 24
1.4 Discourse communities 25
1.5 Multiple discourse communities 28
1.6 Focused discourse communities 29
1.7 Discourse community ownership of genres 30
1.8 Genre action and evolution 32
1.8.1 Emerging order and dynamic change 32
1.8.2 Packaging scientific dialogue – development of the research article 34
1.9 Towards a heuristic definition of genre 36
1.10 A definition of genre and a proposal of 'bridge' genres 40
1.11 A definition of genre with implications for non-native English speaker users 42
1.12 Exploiting genres 44

2. 'Nested' structures, communities and issues 45

2.1 Problems of the 'linguistically challenged' scientist ... 45
2.2 The notion of discourse community 'jurisdiction' 48
2.3 'Nested' structures 50
2.4 'Nested' communities 51
2.4.1 English-speaking community (ESC) 51
2.4.2 English-speaking academic community (ESAC) 52
2.4.3 Focused global discourse community (FGDC) .. 53

2.5 ‘Nested’ issues ... 53
2.5.1 Type I issues ... 54
2.5.2 Type II issues ... 56
2.5.3 Type III issues ... 58
2.6 The notion of a frozen genre ... 60
2.7 Consideration of alternative rhetorics ... 62
2.8 Focus on Japanese rhetoric ... 64
2.8.1 Alternative linguisticultures – a consideration of Japanese ... 67
2.8.2 Vagueness in Japanese ... 69
2.9 Different purposes for language and rhetoric ... 71
2.10 ‘Thought’ without ‘thinking’ ... 74
2.11 The possibility of alternative forms of a genre ... 74
2.12 A need for genre evolution in science discourse ... 76
2.13 Searching for a way to introduce change ... 78

3. The review article and its place in the discourse community ... 81

3.1 General considerations ... 81
3.2 Journal instructions ... 82
3.2.1 Genre name ... 82
3.2.2 Authorship ... 84
3.2.3 Summary of journal expectations ... 85
3.3 Specialist informants ... 86
3.3.1 Rationale for consulting specialist informants ... 86
3.3.2 Specialist informants consulted ... 88
3.3.3 Questionnaire responses from the specialist informants ... 88
3.3.4 Interviews with specialist informants ... 89
3.3.4.1 ‘Explosion of information’ ... 90
3.3.4.2 Different types of reviews ... 90
3.3.4.3 Experts as authors ... 91
3.3.4.4 Cited half-life ... 92
3.4 Definition and propositions for the review article ... 92
3.4.1 Definition of the review article ... 92
3.4.2 An opportune genre for a new age ... 94

3.4.2.1 Influence within the discourse community 94
3.4.2.2 Coping with information explosion 95
3.4.2.3 Bridging discourse communities 96
3.5 Choice of corpus 97
3.5.1 Bibliometric evaluation 99
3.5.2 General description of review articles in the corpus 99
3.6 Preparation of corpus for analyses 100
3.7 Rhetorical organization and physical features of corpus reviews 102
3.8 Classification of reviews based on interviews with specialist informants 109
3.8.1 History reviews 113
3.8.2 Status quo reviews 115
3.8.3 Theory reviews 118
3.8.4 Issue reviews 119
3.9 Type classification of reviews based on text evidence . . . 119

4. Microanalyses of review articles 121

4.1 Introduction 121
4.2 Moves and Steps 125
4.3 Move and discourse organization 128
4.4 Step features 129
4.4.1 SIE – sentence-initial element 129
4.4.2 GS – grammatical subject 132
4.4.3 Vb – verb tense/modality 135
4.5 Step types 136
4.6 SIE features and Step types 137
4.6.1 SIE: Grammatical subject (GS) 138
4.6.2 SIE: Question 140
4.6.3 SIE: Location in space, discourse entity (SpDE) . . 140
4.6.4 SIE: CauseResult (CauseR) 141
4.6.5 SIE: Contrast/concession (Con) 141
4.6.6 SIE: Time Real world entity (TRWE) 142
4.6.7 Summary of SIE findings 142

4.7 Verb tense/modality features of Step types 142
4.7.1 Present tense 143
4.7.2 Past tense 144
4.7.3 Modal verb 145
4.8. Discourse function of the grammatical subject (GS) and Step type 146
4.9. Steps and review types 150
4.9.1 History reviews 154
4.9.2 Status quo reviews 155
4.9.3 Theory reviews 157
4.9.4 Issue reviews 158

5. Discourse organizer and move steps of review articles 161

5.1 Identifying the features 161
5.2 Discourse Organizer Steps 163
5.2.1 Discourse Organizer Step: Topic – Introducing a topic 163
5.2.2 Discourse Organizer Step: Aim – Expressing aim 164
5.2.3 Discourse Organizer Step: Disc (DiscourseSignal) – Organizing discourse 167
5.2.4 Discourse Organizer Step: Sum – Summarizing 171
5.3 Move Steps – Presenting 174
5.3.1 Move Step, presenting: Def – Giving a definition 174
5.3.2 Move Step, presenting: His – Describing historical events 175
5.3.3 Move Step, presenting: Phe – Describing generally accepted phenomena 177
5.3.4 Move Step, presenting: Tech – Describing key technology, method, tool 179
5.3.5 Move Step, presenting: CR – Describing current research 182
5.3.6 Move Step, presenting: Ela – Presenting examples, explanation 185

5.3.7 Move Step, presenting: Lit – Introducing related literature 185
5.3.8 Move Step, presenting: LitH – Introducing historical literature 187
5.4 Move Steps – Evaluating 188
5.4.1 Move Step, evaluating: HisH – Hypothesizing about historical events 188
5.4.2 Move Step, evaluating: AsFL – Assessing findings, literature 189
5.4.3 Move Step, evaluating: Gap – Describing a gap or limitation in the knowledge of the field 192
5.5 Move Steps – Synthesizing 195
5.5.1 Move Step, synthesizing: Hypo – Hypothesizing about specific phenomena 195
5.5.2 Move Step, synthesizing: Dev – Explaining possibilities for further development and applications 198
5.5.3 Move Step, synthesizing: Op – Giving opinion (judgments of value, validity) . . 199
5.6 Summary of Step features 200

6. Review types, moves and steps 207

6.1 Review type classification 207
6.1.1 History Reviews 207
6.1.2 Status quo reviews 210
6.1.3 Theory/model/hypothesis reviews 215
6.1.4 Issue reviews 218
6.2 Components of review articles 221

7. Writing-support concepts and tools 223

7.1 'Publish or perish' 223
7.2 Towards writing support at the lexical level 225
7.3 Towards writing support on the genre text level 227
7.4 'A construction of the learner' 231
7.5 'Negotiated' texts 233

7.6 A writer's 'amanuensis' . . . 235
7.6.1 Help with the lexis . . . 235
7.6.2 Help with discourse community expectations . . . 236
7.7 An amanuensis for review article writing . . . 237
7.7.1 Title and statement of aim . . . 238
7.7.2 Section headings . . . 238
7.7.3 Abstract . . . 238
7.7.4 Review focus and organization . . . 238
7.7.5 Move Steps . . . 238
7.8 Freedom for creative expression . . . 239

8. Towards a more open forum for science . . . 243

8.1 A bridge genre where 'truth comes of age' . . . 243
8.2 The review article – a challenging genre . . . 246
8.3 The review article – an opportune genre . . . 246

Appendix 1 . . . 251

Appendix 2 . . . 255

References . . . 257

Index . . . 271

Preface

The kernel idea for the study described in this volume came from a question from a Japanese scientist who had been invited by an overseas scientific journal to write a review article. He had written many articles for journal publication but this was the first time he had been asked to write a review and had no idea what to do. As I had helped him many times with the English editing of his papers, he turned to me for help. I could not give him a clear answer and therefore embarked on a search for one.

A simplistic solution could have been offered by dealing with surface problems encountered by nonnative speaker of English (NNS) writers, however my experience in helping NNS had indicated that there was much more than merely grammatically correct language at stake in a science text. What started as a simply expressed aim was transformed into an exploration of the foundations of scientific worldviews and the discourses that express them. It led to revealing the hitherto nebulous features of an important genre in science and technology, the review article.

As evident from the scientist's question, these theoretical considerations need to be translated into practical ways to help the NNS. Suggestions as to how this might be done come from technology and engineering fields, such as work in computer science and machine translation. One basic premise in these disciplines is that solutions to problems can be found by breaking down a problem into manageable parts. For example, in order to develop a robot that can walk on two feet, the upright walking process needs to be first analyzed into reproducible components. This approach can also be applied to the work of experts. One example of this is the case of a Japanese company which had to find ways to speed up its mold-designing process to win over competitors. The company, which produces precision molds for electronic equipment such as cellular phones, had their master craftsmen followed by record keepers who precisely noted each step in the planning, designing and production processes. This information was incorporated into a computer-based expert system which drastically cut the mold-designing time. Additional advantages offered by the new computer system were that the three-dimensional on-line designing procedure allowed for quick changes and adaptations which were automatically reflected in the mold dimensions. What is important to note here is that using technology to support a complicated process does not necessarily detract from the creativity expressed in what is to be produced using the support system.

A similar support process should be of great help for the NNS for specific writing tasks using available technology without sacrificing the creativity or self-expression of the writer. In fact, as discovered by the company described above, the technological support can free the designer from concern about the 'mechanical' elements of the process and help promote creativity. Therefore, I draw on ideas from various fields outside of applied linguistics to try to propose practical ways to help the NNS more readily participate in the ongoing discourse of the science community.

The aims of this volume are threefold:

- To consider the theoretical issues involved in applying the notion of discourse community to that of an international science discourse community and examine the issues faced by those who must use a language system that is not their mother tongue in order to become active participants;
- To characterize a genre in which alternative rhetorics might be able to contribute to an international science discourse community – the review article;
- To present practical ways in which writing support can be offered to NNS – concepts from engineering fields are combined with findings from genre analysis and corpus linguistics to propose a framework for helping NNS scientist writers.

The first aim to be addressed is the application of the notion of discourse community to an international academic discourse community in which nonnative speakers of English should be able to play an active role in formulating the ongoing discourse and belief systems. While different rhetorical systems can interfere with mutual understanding when the surface language features reflect unfamiliar information presentation patterns, they may also be able to offer innovative ways of interpreting situations and constructing new ways to view the world.

The next step will be to find some way in which alternative rhetorics can be employed in a manner acceptable to the discourse community. One way this might be done is by identifying a genre in a community which might allow relative freedom of expression and yet contribute toward the shaping of knowledge in that community. The science review article turned out to be such an opportune genre.

The final issue to be dealt with is to find practical ways in which writing support can be offered to NNS based on consideration of discourse community expectations. An approach will be suggested utilizing the features of a set of science review articles and a range of concepts from work in corpus linguistics, phraseology, second language acquisition, computer software and robotics. The support system proposed should be able to offer aid with the more mechanical features of writing in order to promote rather than detract from creative expression.

In order to advance the discussion, four key ideas will be proposed:

- 'Nested' issues and negotiated texts – the issues faced by a non-native speaker of English (NNS) are complex because they represent the expectations of 'nested communities.' No text for a professional discourse community can be written without negotiation with these communities. Presenting these expectations clearly as guidelines and frameworks should help an NNS to cope with these issues.
- An opportune bridge genre – the review article is a genre in the natural sciences which plays an important role bridging a specific discipline and a wider scientific community. Because of this, it offers an opportunity for expanding scientific horizons that is not available to other scientific genres more strictly bound by individual discourse community traditions.
- Moves and Steps – to help an NNS trying to write a review article, key words and phrases to express Steps and Moves are identified. These features are identified by genre analysis and corpus linguistics techniques.
- Frameworks and freedom of expression – concepts such as adaptive imitation and interaction dynamics from robotics research

can offer suggestions for writing support tools to help NNS with writing tasks yet allow for the freedom of expression essential to the production of a creative text.

The work presented here is based on the premise that science is a dynamic way of viewing the world, which is accepted by the community in which it is practiced, and is subject to change by the subset of members who compose the scientific community. This group of people is also influenced by the wider contemporary community in which it spatially and temporally exists. As scientific ideas are constantly changing via interactions among members of the immediate and surrounding communities, the means for sharing these ideas are needed – this will be referred to as the discourse of the community. The importance of these interactions among its members has led to the use of the term 'discourse community' (Bizzell 1982, 1992, Swales 1990a, Berkenkotter et al. 1991). In the scientific discourse communities of today, written texts, and more specifically, English written texts, are extremely important forms of the discourse.

To deal with the complex interactions of a professional discourse community, a range of different types of texts would be needed (Myers 1990b), and these types will be referred to as 'genres' (Swales 1990a, Paltridge 1996). These genres can be expected to have characteristic features at the conceptual, rhetorical and lexical levels which would differ to suit their purpose and their audience. In order to fulfill their purpose, they should offer efficient means of packaging information so that it can be recognized for what it is trying to do and evaluated for acceptance, rejection or further consideration. The people who do the evaluating compose the audience who need to be considered with respect to the explicitness of the language, which can range from a shorthand understandable only to the writer (as a single audience), such as in the case of research notes, to the suitable use of technical terminology when addressing the professional discourse community, and the use of more accessible explanations as the genre type proceeds in the direction of texts for a more general audience. In other words, the genre features are determined by the purpose they serve and the audience they address.

When a genre becomes well established, it can assume an identity of its own. Mauranen (1993a) claims that the genre exists on a higher semiotic plane than those of register and lexicogrammar. In other words, a genre can take on unique characteristics which are more than the simple summation of its register and lexicogrammatical features. This can be compared with living people being more than simple combinations of physical organs and structures. When a genre assumes features which distinguish it from other related genres, it can be used by the members of its discourse community as a means of achieving their "private intentions within the framework of socially recognized purpose(s)" (Bhatia 1993: 13).

As an example of an established genre, there is the science journal article which has been under development by the community of scientists in the physical and natural sciences over the past three hundred years. Today, it embodies the concepts of the scientific method developed for the natural sciences in the West and is now even being used by researchers in the literary and social sciences. Evidence that it is recognized as a well established genre comes from the fact that textbooks have been published to teach how to write a journal research article (Weissberg and Buker 1990, Swales and Feak 1994), and they include examples from various disciplines. In other words, the journal research article seems to have assumed an identity of its own outside of the natural science community in which it was originally developed and is now being used as a tool to examine issues in other communities.

This genre of the research journal article has also been extensively studied by those interested in the history of science as well as applied linguistics and second language pedagogy (Bazerman 1991). The textbooks mentioned above are aimed at teaching nonnative English speakers how to write it in order to enable them to gain a voice in their international discourse community. Good research would be wasted if it could not be offered for evaluation by other professionals working in the same field. In today's world, this presentation must be made available to an international audience with the language of choice usually being English.

This book will examine a genre that has been less well studied – the review article – but is nevertheless very important in the natural sciences. There has been some work done by Myers (1990b, 1991), but the review article has not been subjected to much linguistic scrutiny (Swales 2003). I found that the term itself was often mistaken for 'book review article' by people in applied linguistics. Hereafter in this volume, 'review article' will refer to an article presenting an overview of a field of professional research in the natural sciences or of some important element of it, usually not emphasizing original research by the author(s).

The review article will be revealed to be a genre in the natural science discourse community that can serve as a bridge between the discourse community of a specific research area and the overall scientific discourse community. These claims are supported by analyses of a corpus of 25 review articles from a prestigious English-language science journal, *Proceedings of the National Academy of Sciences of the United States of America,* and evidence from expert informants. The review article is shown to be a forum for the synthesis of ideas which can be presented to those outside the immediate discourse community. It seems to be a formal debut genre of 'a scientific fact' – this is where a 'fact' can be presented to the wider sphere of a more general academic community.

The ultimate aim of this study was to suggest how to help non-native speakers of English (NNS) write review articles in order to gain a voice in the dynamic forum of scientific rhetoric. This required an examination of second-language issues and a categorization of the various genre features in order to find feasible ways to offer support. Useful concepts are employed from fields of corpus linguistics, second language pedagogy, machine translation and robotics to propose a system of writing support tools based on adaptive imitation and interaction dynamics. This is done with the hope that such tools will be able to provide the linguistic, rhetorical and cultural support to allow scientists from non-Anglo-American backgrounds gain a respected voice in their discourse communities. Also, as the review article genre seems to offer more expressive freedom than other scientific genres and is usually commissioned by journal editors, it may well serve as the means for awakening the consciousness of the scientific community to new vistas via the introduction of alternative rhetorics.

This study began with a specific question and ended with a specific answer but the search for that answer took a meandering path through ideas about language itself and its place in science, or a way of viewing the world around us. The path finally led to one way of categorizing review articles and an understanding about their place in the construction of 'science' as it is practiced today and a synthesis of concepts from diverse fields. Although the 'answer' comes too late to help the person who asked the original question, I hope my reply can be useful for those who would like to know about the review article in its present form and some practical ways to address the issues involved in second-language writing for a discourse community in order to allow them to have a voice in constructing the dynamic worldview of science.

My deepest appreciation goes to Tony Dudley-Evans and Susan Hunston of the University of Birmingham for being the guiding lights for this research and to John Swales of the University of Michigan for his encouragement and interest (any mistaken paths are of my own choosing). I also express my gratitude to the scientists who so generously allotted me time from their busy schedules to act as specialist informants.

I thank the National Academy of Sciences, U.S.A., for kindly granting permission to quote from the reviews which were analyzed in this study. The full journal reference information for each review is presented in Appendix 1. ("Permission granted" fax transmission dated 04/29/04 from Diane M. Sullenberger, Executive Editor.)

I am very grateful to the guidance from the *Linguistics Insights Series* Editor, Maurizio Gotti, and to Peter Lang AG editors Monika Lehner, Silvia Mueller and Vera Reid.

I dedicate this work to Eisuke Noguchi who makes everything possible with his constant optimism, understanding and support.

1. Discourse Communities and Scientific Genres

The truths of science, then, are achievements of argument.
(Gross 1990: 103)

1.1 "No language, no facts"

Science and rhetoric cannot do without the other, and both need a receptive audience. Gross (1990: 203) states that "facts are by nature linguistic – no language, no facts". And together with language comes the notion of how it is used, or rhetoric. Thus, we have the connection of science and rhetoric, which has been discussed by many (Woolgar 1976, Fleck 1979, Lynch 1982, Law and Williams 1982, Gilbert and Mulkay 1982, Stonier 1984, Myers 1985, 1986, 1988, Latour and Woolgar 1986, Harwood 1986, Fahnestock 1986, Brufee 1986, Bazerman 1988, Zappen 1991).

Thinking about this relationship among science, rhetoric and society can be traced back to the introduction of Kuhn's theory of scientific revolutions (1962), which was followed by a blossoming of studies in the cognitive sociology of science as reviewed by Knorr Cetina (1981: 22) and Collins (1982: 44). In 1982, Collins wrote about "tacit knowledge and scientific networks" and discussed "social circles" of scientists which "is 'a fuzzy edged' group whose members associate more with each other than outsiders in respect of one or more social relation" (Collins 1982: 61). He cites an "impressive" number of studies published from 1969 to 1972 related to the "invisible college" idea of Price (1963), but berates them for having missed the point of actor participation in their preoccupation with observing only the information via bibliometric indices and questionnaires. Collins refers to Burns (1969) having written that "technological knowledge is the property of people rather than documents" (Collins 1982: 45). In his conclusion, Collins remains dissatisfied with the research methods in use at the time but his paper contains the seeds of the flowering that was about to come as can be seen from the ample list of references from the 1980's.

In 1990, Myers tries to capture almost "all the elements and uses of scientific knowledge, not only the formal and informal elements, but the political, persuasive, and emotive, and even the intangible and unspeakable" that Collins considers necessary to cover – to even "speak *about* that which cannot be spoken" (Collins 1982: 61 [emphasis in original]). Myers passionately writes in the Preface to his book *Writing Biology: Texts in the Social Construction of Scientific Knowledge* (1990a: xii):

> By including the word biology in the title, I hope to get this book placed on the rapidly lengthening shelf of studies of science, and particularly with studies that do not necessarily take physics as the typical science. I hope it will become clear, after the first chapter, that the title is not Writing ABOUT Biology because that would imply that biology is there before the writing and that the writing merely dresses it up. I argue instead that writing produces biology. [emphasis in original]

Myers (1990b) traces how a scientific discovery is not reported as such at first, but appears merely as a claim. It only becomes a 'discovery' after being retold as an event when it is accepted by the discipline / discourse community:

> I am going to argue that the attribution of discovery is made, not just in texts, but between texts, in the interpretation of one text by another, as a process of reading as well as writing. This process of interpretation is an essential part of the processes of recording, negotiating, selecting, arranging, and transforming statements, the processes that produce scientific knowledge. So if we want to find a discovery, we can't just go back to the original articles or even earlier to the lab notebooks or recorded conversations or autoradiographs and electron micrographs. We need to look at the interpretations of the articles as their stories are retold in news articles, review articles, textbooks, and popularizations. (Myers 1990b: 102-103)

The general public only becomes aware of the 'discovery' when it has been recognized by the discipline / discourse community and has become "frozen into a fact" (Myers 1990b: 114). The struggle which had been waged until then is no longer needed because the 'discovery' has won its place in the community consciousness.

This seemingly dispassionate success entrances not only the general public but the scientists themselves. Humphries, formerly an 'insider' with formal training in physics and biology, who long worked on research related to environmental issues, warns that: "scientists falsely assume that science is objective. It has its internal consistent logic, but in the context of the choices scientists make to do their work, it is highly subjective" (Humphries 1998: 9). A visiting scholar at the Sloan School of Management at the Massachusetts Institute of Technology, Humphries feels the need for more acknowledgement of this subjectivity and a movement away from "disembodied" writing which can create a false illusion of what is really an opinion assuming the guise of an objective fact. Humphries quotes from a *Science* article, 'Cut the Communications Fog', which points out the "obscure, techie writing" which seems to be plaguing the scientific discourse today. She states that this "disembodied voice led to the loss of accountability, relevance and meaning, and is a reflection of the 'deeper systemic issues' of society". Disturbed by the way in which scientists were "confined to the narrow window of giving and interpreting technical knowledge", she left the field when asked to participate in an international project to develop biodiversity assessment tools for developing countries. This was because she realized that the scientists were being used to grant an "imprimatur" for land development. Before leaving science, she wrote a book using a style which clearly expressed her stand on ecological devastation which could be caused by outdated land management policies in Australia and, thus, was able to influence national policy in related areas. She has since been involved in work on social issues.

To the general public, science presents an image of being dispassionate and objective, but Gross claims that this is actually an achievement of its success in persuasion: "My rhetorical analyses show how the sciences construct their specialized rhetorics from a common heritage of persuasion. By means of these, the sciences create bodies of knowledge so persuasive as to seem unrhetorical – to seem, simply, the way the world is" (Gross 1990: 206-207).

1.2 Different audiences, different sciences

Let us continue this discussion of different views of science. Myers (1990a: 248) says: "But if, as I have suggested, different audiences get different narratives [e.g., of a research article for a specialized audience vs. a popularized version for a general audience], and different narratives carry different views of the work of science, those who take a realist and empiricist view of science can only ask which narrative is correct".

This idea of there being a range of versions for different audiences is not limited to studies in applied linguistics. It has also been expressed by Nakayama, a Japanese scholar of the philosophies of the East and the West (Nakayama 1984: xi):

> Thus the notion that science is Greek science and all those modes of learning that do not conform to the Greek pattern are not science would view all such activity in terms of and through its perception of the Greek 'scientific edifice.' When intellectual activity is understood in terms of the acts of creation, development, and transmission, however, it emerges as a human actvity [sic] that could and in fact has occurred everywhere. Thus it becomes possible to speak of Chinese science, Indian science, and even Hottentot science.

In a footnote, Nakayama comments on his use of the word 'science' (Nakayama 1984: xi):

> In this book I have avoided any hard and fast definition of science. From my point of view, it is not very meaningful to take the science of a particular period or region and make it into a kind of ideal type.

If we accept this idea of 'intellectual activity' as being 'human activity,' then the next step would be to see the possibilities of there being different bases from which these 'acts of creation' could occur, develop and be transmitted. Projecting this to the field of science, Nakayama states that "As an East Asian, I cannot be persuaded that the course of scientific development followed by the West is the only logically necessary course" (Nakayama 1984: xvi). This is after he has defined 'the West' and 'the East' as follows:

> By the West, I mean in effect the historical mainstream of scientific development. Though there were frequent shifts in its historical center [...] this is the stream that eventually grew into contemporary science. By the East, I mean science as it was practiced in the Chinese cultural sphere, in almost complete isolation from the mainstream. (Nakayama 1984: xiii)

What I wish to suggest here is that there can be different narratives from different cultural communities which can contribute to global scientific developments. For example, the documentary style of learning often associated with ancient China (Nakayama 1984: 3) differs from the rhetorical scholarship tradition that began in ancient Greece (Nakayama 1984: 6) and laid the basis for the Western approach to science. However, both can contribute to scientific development. What is important today is that these different communities all become aware of what is taking place in the sciences and, if possible, play more active roles in the creation of their narratives. Myers notes that "scientists have neither been surprised nor felt threatened by my comments on scientific rhetoric" (1990a: 248), and calls for more involvement of the general public in the development of science:

> Although I do not see this book [examining the role that texts play in the social construction of scientific knowledge] as a debunking project, I do see the analysis of scientific texts as a way of promoting change. Specifically, I would like nonscientists to read more science, to read it more critically, to read it with an awareness of the social processes that produce it, and to question the authority with which science is sometimes presented in cultural and political contexts. (Myers 1990a: 249)

Of course, this definition of science as being a world-view developed through rhetorical discourse is not to "denigrate" it at all, but "rather to pay tribute to the supreme human achievement that consensus on complex issues represents" (Gross 1990: 103).

1.3 From chaos, new gestalts

These ideas of discoveries arising out of a complex web of "recording, negotiating, selecting, arranging, and transforming statements" with the process of interpretation being essential to progression (Myers 1990b: 102-3) agrees with how humans try to make sense out of what seems like a chaos of possibilities. We humans try to make sense of this world we are in by representing it through science. And science itself continues to reveal more and more about how human cognition seems to be shaped. From bioscience research over the past 30 years or so, Freeman (1991) and his neurobiology group at the University of California, Berkeley, have been trying to discover how people can instantaneously recognize a famous actor's face, a friend's voice, or the mouth-watering smell of a favorite dish; what psychologists call 'preattentive perception.' In studies on the physiology of olfactory perception, they have found that consciousness may result from a "recursive process of motor command, reafference and perception" (Freeman 1991: 41), which is governed by chaotic, collective activity. In other words, when the brain detects stimuli in the vicinity, it sends out a motor command to have the person receive more sensory information. The neurons in the region involved go into chaotic, collective activity to form a gestalt which is continually adjusted by new information. Chaos is "complex behavior which seems random but actually has some hidden order" (Freeman 1991: 34) and makes possible the production of novel activity patterns which Freeman's group proposes to be "crucial to the development of nerve cell assemblies that differ from established assemblies." In other words, it makes possible insight and problem solving through trial and error. Chaos seems to be the very feature that makes possible instantaneous perception and also makes the brain distinct from an artificial-intelligence machine, which cannot create new gestalts through such chaotic activity.

This repeated confrontation with chaos and the formation of new gestalts including new information are also activities engaged in by the discourse community. Here I would suggest that the chaos of new information is perceived by the community and the gestalts that are formed compose the 'science' of the current community. When an individual forms a gestalt in reaction to stimuli, it results in the recognition of a face or a smell based on sensory signals. When a community forms a gestalt, it results in the recognition or understanding of a phenomenon based on the signals which are transmitted among the community members by language.

This language itself is also in a 'chaotic' state, i.e. it is not stable. Halliday and Martin describe Lemke's characterization of a language as a dynamic open system (Halliday and Martin 1993: 15):

> a system that is not stable, but is metastable, able to persist through time only by constantly changing in the course of interacting with its environment. One way in which a language typically changes is that new registers or functional varieties evolve along with changing historical conditions. The evolution of a register of science is a paradigm example of this.

What Halliday and Martin are saying seem to be analogous to the discoveries about human cognition – the science that we believe in (which is an extension of human cognition) at any certain time is a gestalt that has been arrived at by the members of the professional community. Out of a chaos of possible views, the community at any certain time accepts a set of beliefs that informs their ways of interpreting their experiences and directs their actions. In the case of the brain, nerve cells send out electrical signals to direct interaction with the environment. In the case of a discourse community, the signals are sent out through the medium of language.

1.4 Discourse communities

As communication has become possible over greater distances with almost real-time access, people can form groups that are not necessarily in physical proximity. For example, in today's networked world, it is simple for a scientist in Japan to be in constant contact with scientists in England or the United States or China or anywhere else covered by the communications web. Therefore, polymer chemists who specialize in developing polymer microspheres of unusual shapes for special purposes, such as drug delivery systems or inks for computer peripherals, often find themselves more closely allied scientifically with scientists doing similar work in Europe or Korea than someone working on the development of polymer films in a physically adjacent laboratory. I became aware of this 'absence' or 'incomprehensibility' of professional disciplinary communication when trying to teach graduate students in the same department working in adjoining laboratories. Both groups were in polymer science, but unable to understand each others' papers because of the differences in content background. In other words, the polymer scientists studying polymer microspheres belong to a community of polymer chemists specializing in polymer microspheres which differs from the community to which those studying polymer thin films belong.

The origin of the term 'discourse community' as it is being used here is difficult to define. Swales (1998: 197), reminiscing in a section on "The concept of discourse community revisited" states that he first heard this term in early 1986 and "adopted it immediately as it captured and crystallized" what he had been thinking about at the time.

Since then, much work has been done in trying to capture this elusive concept [Swales lists 20 works published between 1986 and 1997 which deal with it, (1998: 197)]. His 1990 work, *Genre Analysis: English in Academic and Research Settings*, presents a definition of discourse community which still remains perhaps one of the easiest to grasp (Swales 1990: 24-27):

a. having a broadly agreed set of common public goals
b. having mechanisms of intercommunication among its members
c. using its participatory mechanisms primarily to provide information and feedback
d. utilizing and hence possessing one or more genres in the communicative furtherance of its aims
e. having acquired some specific lexis
f. having a threshold level of members with a suitable degree of relevant contacts and discoursal expertise.

More recently, Swales (1998) raises the question of there being a possible difference between a 'discourse' community and a 'disciplinary' community: "Those who opt for *disciplinary community* can be seen as centering on the 'invisible college' idea of a dispersed group of like-minded specialists or enthusiasts, whereas the preference for *discourse community* may focus equally on a local grouping. (I will leave it for others to decide whether this distinction is eroded by the rapid advance of electronic intercommunications.)" Swales (1998: 198). The comment in parentheses suggests that Swales is pointing to physical proximity as a possible characteristic feature.

Swales' comment in parentheses is indeed what has been happening. However, the comment also captures the current growth being witnessed in Web communications, with real-time communications becoming not merely possible, but economically feasible. In fact, there are probably almost no academic conferences of significance which can now be launched without an on-line call for papers and a mailing list of committee members involved in the planning. This ease of communication almost without regard for distance or time tends to promote strong ties among those with similar interests.

Therefore, returning to the polymer scientists described above, we can think of them as belonging to a discourse community of polymer scientists, a larger discourse community of chemical engineers and to still larger communities of scientists and of academics. This would fit the description of 'global discourse communities' tendered by Killingsworth and Gilbertson (1992) described in Swales' book (1998: 201): "*Global discourse communities* [...] are groups of writers and readers defined exclusively by a commitment to particular kinds of action and discourse, regardless of where and with whom they work" (Killingsworth and Gilbertson 1992: 162). They continue, stating that "membership in global communities tends to be regulated exclusively by discourse-governed criteria (writing style, publication in certain journals, presentations at national conventions, professional correspondence, and so forth)" (Killingsworth and Gilbertson 1992: 169).

Swales points out that Killingsworth and Gilbertson originally based their discussion on the work of James Porter (1986) who offers the notion of a 'forum' as "a concrete, local manifestation of the operation of a discourse community" (Porter 1992: 107) (in Swales 1998: 200). It seems that Porter does not limit the 'local' to physical location, or place but includes conferences and journals as 'fora', And Swales (1998: 201) points out that, in juxtaposition to 'place' discourse community, there could be the idea of 'focus' discourse community. Swales then goes on to define a 'place discourse community (PDC)' to test whether his 'small university building' does indeed have a PDC on each floor (and decides that two qualify but one does not). What I would like to suggest here is that the polymer scientist belongs to a 'focused global discourse community', or a community of polymer scientists who are not in the same locality physically but in their professional focus. The word 'global' represents this absence of a need for physical proximity.

1.5 Multiple discourse communities

The possibility thus arises of there being multiple discourse communities, and with them language issues because the predominant language used by the global community may differ from the native language of the individual members. Mauranen (1993a: 5) presents an interesting comment on types of communities: "It thus seems warranted to assume that at the intersection of an international academic community and a national language community there is a national academic community, which exerts an influence on its members and their writing".

She states this after having noted that "research papers cannot be written in Finnish (or Swedish) for an international audience, a great proportion of academic texts are published in English in Finland even when they are for the most part intended for a domestic audience (e.g. forestry or agricultural reports). The motivation is a potential international readership, and the higher status of an international publication compared to a domestic one." (Mauranen 1993a: 3). The same phenomenon can be said to exist in Japan where, for example, in the area of plant biosciences, there is now almost no Japanese-language journal where original research can be published. Even the journals published by Japanese academic societies are English-language ones (personal communication, Masayoshi Maeshima, Nagoya University). This means that participation in the discourse community both globally and locally no longer offers a choice of language, at least for these written reports. Trying to identify an "international scientific

English", Wood (2001: 83) calls for a recognition of English for science and technology as "not something extra, grafted onto training to be a scientist, which is how it is often traditionally seen by students of science, but an integral part of that training".

Of course, there are various forms of communication, such as reviews, correspondence, and oral lectures and discussions aimed at local (geographical proximity) discourse communities. In the case of the polymer scientists described above, the question arises of what features can be said to describe the genres / language(s) used by polymer scientists in communicating with the various groups of which they are members. American or British scientists are likely to use English for all of their groups but scientists whose native tongue is not English will probably use the vernacular. In the case of Japanese scientists discussing science in a discourse community within Japan, they would most likely be using Japanese, but they would still be functioning within a scientific discourse community. Thus, some features of their language would be academic, scientific and more specifically, peculiar to the polymer science discourse community. This could explain why a native Japanese speaker who did not know anything about an academic discourse community, much less polymer science, would not be able to understand or participate in such an interchange.

1.6 Focused discourse communities

Projecting from Mauranen's (1993a) comments on Finns writing for differing audiences, I would argue that people in professions participate simultaneously in the ongoing discourse of several communities. At least, in the case of the Japanese polymer scientists mentioned above, there would be a focused global discourse community in which the communications would be in English and a focused local discourse community in which the local vernacular would prevail as the mode for communication.

Based on consideration of Swales' discussion of Killingsworth and Gilbertson and Porter, the notion of a 'focused global discourse community' (FGDC) is used in this volume. An FGDC would be a group of people who are in a dynamic state of equilibrium in relation to disciplinary cognitive issues and have mutually recognized means of communication by which they endeavor to maintain this dynamic state and also allow it to evolve. The 'focused' reflects the attention to a specific 'cognitive issue'. 'Global' means that the community is not bound by physical restraints. 'Discourse' refers to the fact that the established means of communication among the members is crucial to the existence of this entity. The word 'community', of course, reflects the fact that people are the constituents. Another aspect of the notion of community is Swales' observation (1998) that a functioning community will also have defense mechanisms to resist external intervention or change.

1.7 Discourse community ownership of genres

This brings us back to the language which the community uses for their interactions. As in the quote by Gross (1990: 103) that begins this chapter, "the truths of science, then, are achievements of argument". The existence of 'argument' means that there is a group of people with whom to 'argue' and this group is the focused discourse community. This discourse community, over time, is likely to develop patterns for efficiently expressing these arguments. As these patterns become established, they become accepted by the members as 'genres.' Swales refers to the relationship between the discourse community and a genre at the end of his textography book (1998: 207) and points out that the "prevailing" position (proposed by himself in his 1990 work and by Berkenkotter and Huckin in 1995) is that it is one of ownership; i.e., the discourse community owns the genre. However, he acknowledges the opposing position of Mauranen (1993b), which is that it is the genres that own the discourse community, and the position of Devitt (1996), who points to a need to examine the very relationship between 'genre' and 'discourse community.' He also presents support for Mauranen's position in that the "hallowed genres" of Systematic Botany, which existed before the present members of the PDC entered it will probably remain in existence long after those people are no longer active in this PDC. This discussion suggests that the temporal element and, thus, change over time or 'evolution', is something which should be taken into consideration when examining discourse communities and genres. This notion of genre evolution will be discussed further below. What should be noted here is the concept of a discourse community possessing genres which, in turn, can be thought of as a dynamic binding characteristic of the community itself. Without these genres, there may be no discourse community and vice versa.

As a genre is used for communication, it can be assumed to appear within the discourse community in the form of a text (this term is used here in its widest sense to include oral as well as written forms). Study of such texts could be a key to understanding that community. The respected Japanese philosopher Nakamura talks of "cherished" texts (Nakamura 1964: 9-10):

> [...] in studying the ways of thinking of a given people, one should consider the ways of thinking adopted by most of the members of that group. [...] the characteristic popular sayings, proverbs, songs, mythology, and folklore of that people. In addition, such things as myths, religious scriptures, the arts [...] and works of literature in general must, of course, be considered as important sources for our study. Since such documents abound among all peoples, one must choose as source material those which are *particularly cherished by the people in question*. Those works which are not esteemed by the people [...] will be of little significance as sources for determining the ways of thinking common to the entire people. [emphasis mine]

Nakamura is describing a 'people' such as the Chinese or the Japanese, but what are the Chinese or the Japanese but groups of people who have lived together for long periods and therefore gained a name for their group. When people mingle and form communities, they proceed to organize themselves into subgroups within their individual communities. These groups can be according to trade or interest. What is essential among these groups is communication by which they can reaffirm their reasons for belonging to a certain group.

Nakamura (1964) says that "language is basic to the cultural life of a people" and the same can be said here for a discourse community. If this is so, then Nakamura's approach of studying the texts "cherished" by the people would, in the case of a discourse community, be a study of the texts considered important by the members. In the case of the scientific discourse community, these cherished texts could be items such as the lab notebook, the conference abstract, the research journal article, and the review article.

Of course, Bakhtin (Holquist 1981) would probably say that "cherished" texts are fossilized and therefore not as interesting to study as something such as the novel, which is still under development. However, rather than being counterevidence, this suggests that genres can be at different stages of development – those in creation, those under development, those evolving into new forms, and those which have stabilized.

1.8 Genre action and evolution

Before going into a detailed analysis of genre features, let us consider genre action and the evolution of genres because they will become central to one of the purposes of this volume – making the genre features of a prominently Anglophone global discourse community more accessible to a non-native speaker participant. As stated above, the notion of a focused global discourse community (FGDC) is being used here to refer to a group of people in a dynamic state of equilibrium in relation to disciplinary cognitive issues. The means by which they maintain and develop this state is language packaged in the form of mutually recognized genres. As the FGDC is not static but dynamic, the genres themselves are also dynamic, meaning that they are in a continual state of evolution.

1.8.1 Emerging order and dynamic change

To try to understand why evolution itself occurs, work with mathematical models and biology is illuminating. Biologists are starting to apply these models in trying to explain natural selection as arising from antichaos and adaptation. Kauffman, professor of biochemistry and biophysics at the University of Pennsylvania, School of Medicine, suggests that random Boolean networks, used in computer models, may help explain biological development and evolution (1991). The behavior of complex systems, such as those found in biology, is affected by innumerable factors. Boolean networks are used to try to express the dynamic behavior of these variables. The results indicate that order emerges spontaneously, i. e., almost all living things display homeostasis, or the "maintenance of an internal consistency and an independence of the environment" [McGraw-Hill Dictionary of Scientific and Technical Terms, 5th ed. (Parker 1994)]. This state is marked by the existence of "a frozen core" which creates constancy. Around it are "islands of changing elements." Evolution occurs at the edge of chaos. This work is being applied to the study of biological evolution, with reasonable results. It enables calculation of the number of cell types, and thus, prediction of the complexity of the organism based on the number of genes in its genome, or the entire set of genes of the organism – jellyfish are more complex than sponges but less complex than fruit flies. Humans are the top-of-the-line product.

Let us try to apply this argument to genre evolution. In the social sciences, Knorr Cetina, in trying to define the use of metaphors in science (an interesting example of which is employed by Kauffman above) states that in society today, the focus is on the study of institutions. Citing Douglas (1966), Knorr Cetina goes on to say that as institutions are "conventions" or "socially contrived arrangements" (Knorr Cetina 1995: 331), they require some form of justification for their existence. This justification is satisfied by links with "'natural categories' through metaphor and analogy" (Knorr Cetina 1995: 331). The idea of 'natural categories' goes back to work in the 20th century on primitive classifications which were "symbolic classifications of natural objects by native tribes" (Knorr Cetina 1995: 330).

Berkenkotter and Huckin (1995) discuss the concept of genre change by pointing to how background knowledge of the community can affect genre content. They describe the results of a study showing how news reports changed over a 40-year period to accommodate shifts in community assumptions. Another example is that related to the *kairos* concept expressed by Miller (1992), who points to differences in two papers related to early work on DNA, separated by a nine-year time span, which seem to belong to different genres. Berkenkotter and Huckin point out that "scholarly articles are expected to be at the 'cutting edge,' to make an original and novel contribution to disciplinary knowledge", and this newsworthiness feature is important to properly "orchestrate" the review process (Berkenkotter and Huckin 1995: 15-16).

Thus, in both the natural sciences and social sciences, dynamic objects or institutions are seen to come into existence and evolve. This can help explain why genres form and develop: when an entity is forming, there is a need for the "maintenance of an internal consistency and an independence of the environment".

1.8.2 Packaging scientific dialogue – development of the research article

From a chaos of many possible forms and formats, discourse communities have evolved genres to be able to elicit "reasonable results" from many possible ones. For example, in order to develop lines of communication that can efficiently convey the kinds of information essential to their purposes, the scientific discourse community has developed the research article to describe experimental research. The first Secretary of the Royal Society in England, Henry Oldenburg, became the first scientific editor with the establishment of the *Philosophical Transactions* of the Society (Gross 1990: 176):

> In 1683 the Society passed a resolution to emphasize the importance of record keeping and putting things in writing (Birch 1968, IV, 251). By 1686 there were clear signs of a regular procedure [...] the trend toward dated journal publication as a guarantee of temporal priority soon took full hold: by publishing in print accounts of their completed discoveries or inventions, scientists simultaneously shared their results and established their claim to these results as theirs alone. By creating the means by which the ownership rights to its knowledge could be established, the scientific community transformed rhetoric into social reality.

This establishment of procedure and ownership rights can be thought of as ways of maintaining internal consistency and gaining independence from the environment. Also, as discussed above, the discourse community may own a genre but the genre itself, once established as a means of communication among its members, can contribute to change within the community.

This description of the events occurring in the 17th century at the Royal Society parallels the descriptions of Kauffman (1991) and Knorr Cetina (1995) about the formation of loose nebulous entities which gradually develop 'core' structures or 'conventions'. In the case of the Royal Society, the simple letters sent to Oldenburg were gradually incorporated into a 'regular procedure', which in turn, through this sharing of 'knowledge', created a shared consciousness among the members of the Society. In other words, 'rhetoric', or language, by being shared via 'recognized' forms among a specific community of people, had been transformed into a 'science,' or a way of viewing the world.

Thus began the development of the research article as we know it today. Bazerman (1988) describes the experimental report as having "a ubiquity that seems to overshadow the others [other genres]. The experimental report seems central to many conceptions of the sciences as empirical inquiry. The experimental report has developed as a favored solution of the problem of how to present empirical experience as more than brute fact, as a mediated statement of inquiry and knowledge" (Bazerman 1988: 7).

The genre of the experimental report, or research article, has now emerged from a chaos of possibilities as a 'favored solution'. Natural complex systems, as described by Kauffman, demonstrate a tendency to develop orderliness (1991). In discourse communities which are 'natural' because humans are part of the natural world, the forms of intercommunication spontaneously tend toward an ordering because this can raise the efficiency of the system. After all, as Miller states "successful communication would require that the participants share common types" (1984: 157). Of course, if the type becomes too rigid, then development ceases and as with any living system, this would signal death. Therefore, there is a need for the system to remain in a dynamic equilibrium with continual adaptation to the ambient environment. Only in this way can it remain viable and productive. What I would like to propose here is that it is at this chaotic evolutionary edge of the community that genres are evolving or emerging. This chaotic edge, far from being a negative factor, is in fact a window of opportunity where there can be input from other cultures, other ways of viewing the world, other gestalts in order to help global discourse communities develop and cope with continually arising new issues.

1.9 Towards a heuristic definition of genre

One of the main aims of this volume is to establish that the review article is an important genre in natural sciences. In order to do this, 'genre' must first be defined, and then a decision made as to whether the review article satisfies this definition. The definition of 'genre' will be based on the concept of science as a world view achieved by consensus in the scientific discourse community, which was discussed above. A community consensus can only be reached through interactive communication that can take many forms – formal to informal, written and spoken. As the spoken forms are difficult to capture and analyze, there has not been much work done on them, except perhaps the ethnographic type done by Knorr Cetina (1981) and more recently, Swales in his bold venture into textography (1998). However, it is still a given that knowing the written forms "in one's field is [...] essential to professional success" (Berkenkotter and Huckin 1995: 1). Therefore, the focus here will be on the written forms, which is in line with the main focus of this volume – the review article.

In applied linguistics, the concept of genre was first expressed by the word 'register' in 1956 by Reid and further developed by Ure and Ellis (1977) according to the history of 'register' described by Leckie-Tarry (1995). In 1964, Halliday, McIntosh and Strevens (1964) stated that 'register' is "a variety of language distinguished according to its use." In other words, what a language is used for affects the form of the language itself. Martin, in 1984, defines genre as "a staged, goal-oriented, purposeful activity in which speakers engage as members of our culture" (1984: 25). Thus, this definition of 'genre' now included, in addition to language, the notion of different parts (stages) and a target (goal-oriented) as well as the participants in the activity (speakers and members of the culture).

At about the same time as these developments, much work was being done in the area of English for specific purposes (ESP), or at that time, English for science and technology. Swales (1985) traces the development of ESP by picking out episodes that helped propel the field forward from the 1960's to the early 1980's. The early work generally focused on the grammatical-rhetorical analysis of science texts, e.g., Lackstrom, Selinker and Trimble (1972) investigated how grammatical choices, such as verb tenses and article usage, were related to rhetorical functions in written English for science and technology. However, these analyses were sometimes inadequate and caused some misgeneralizations, such as that pointed out by Swales in 1981. He noted that although some textbooks for English for science and technology emphasized practice in using the definition statement, this structure did not play such an important role in 'real' scientific writing. Definition statements are common in textbooks and school examinations, but not in the texts that practicing scientists read.

Although much work was also being done during this period in discourse analysis (Sinclair and Coulthard 1975, Winter 1977, Hoey 1979, Tadros 1981), as Bhatia points out, there was "no adequate information about the rationale underlying various discourse-types, insufficient explanation of socio-cultural, institutional and organization constraints and expectations that influence the nature of a particular discourse genre" (Bhatia 1993: 10) and there had been "little attention to conventionalized regularities in organization of various communicative events" (Bhatia 1993: 10).

More promising was work on viewing the writer and the target audience in addition to the easily recognizable packaging or encapsulation of information. From the viewpoint of systemic functional linguistics, Kress and Threadgold state that "genres are not simply schemas or frames for action. They involve, always, characteristic ways of 'text-making' (mode), and characteristic sets of interpersonal relationships and meanings" (1988: 216).

It must be noted here, as also discussed by Leckie-Tarry (1995), that Halliday himself does not use the term 'genre' but rather chooses the word 'register' to "encapsulate the relationship between texts and social process [...] he employs 'genre' in a more limited sense [...] genre is a lower order concept, register the higher order concept subsuming genre" (Leckie-Tarry 1995: 7). Halliday's term for the organizational structure of a text is 'generic structure'. Hasan, on the other hand, does use the term 'genre' in Halliday and Hasan (1985: 62) and states that it is defined "by obligatory elements in structure". She "abandons" the traditional usage of genre to describe literature and examines examples of conversation, or everyday situations in which language acts as an instrument (1985: 54). After describing the elements obligatory and optional to a text, Hasan proposes the notion of genre structure potential, which expresses these elements, their ordering and the possibility of their occurrence.

Yunick (1997: 328) sums up and compares the constructs of register and genre noting that "purpose is the organizing concept of genre analysis, but only a component (albeit an important one) in Hallidayan register". He also clarifies the differing emphases among the three major schools of genre analysis today:

> With respect to drawing relations between language and its social functions, New Rhetoric [e.g. Bazerman 1988, Bizzell 1992, Berkenkotter and Huckin 1995] focuses on the social purposes end of the spectrum, while genre analysis in ESP [e.g. Swales 1990a, Bhatia 1993, Johns and Dudley-Evans 1991, Johns 1993] brings more focus to moves in discourse structure. Australian linguistics [e.g. Christie 1994, Martin 1984, Halliday and Martin 1993] explicitly and theoretically hooks up grammar and lexicon as well as discourse structure to social function. (Yunick 1997: 322)

In the model being developed in this volume, the discourse community owns the genre, which can find expression in the form of linguistic texts (there can be other 'texts' such as graphic and performing arts) possessing characteristic features. Because the texts are linguistically expressed, they predictably possess language features, i. e., those of grammar, lexicon, and discourse structure.

While trying to develop a view of the science review article, I am also trying to develop ways of making professional genres more accessible to non-native speakers of English to enable them to participate in the professional dialogues of their communities. Thus, while recognizing social purpose as of primary importance, there is also the need to include both the moves and the 'hook-up' between the grammar, lexicon and discourse structure to reach a realistically useful definition of genre. One suitable starting point is Martin's (1984) definition of genre as "a staged, goal-oriented, purposeful activity in which speakers engage as members of our culture". Mauranen (1993a) notes that the word "staged" means that the activity is structured and, in further developing Ventola's ideas (1987), she refers to genres as being activities on a semiotic plane of communication that is above the level of register or lexicogrammar.

Combining discourse analysis with a model of socio-cultural, institutional and organizational explanations was offered by Swales in his work on genre analysis in 1981 and in a book devoted to it in 1990. He tried to capture the intimate relationship between language and context. Swales' definition of genre (1981, 1985, 1990a) is summarized by Bhatia in his book on *Analysing Genres* (1993: 13):

> a recognizable communicative event characterized by a set of communicative purpose(s) identified and mutually understood by the members of the professional or academic community in which it regularly occurs. Most often it is highly structured and conventionalized with constraints on allowable contributions in terms of their intent, positioning, form and functional value. These constraints, however, are often exploited by the expert members of the discourse community to achieve private intentions within the framework of socially recognized purpose(s).

Bazerman places genre in the realm of social phenomena which emerges as a strategy repeatedly used to achieve solutions in similar contexts and thus "gain institutional force" (1988: 8). This connection of genre and social force can be traced to Swales and further back to Carolyn Miller. In developing his definition, Swales gained insights from the work of Miller (1984) in 'Genre as Social Action'. In Bazerman's book of 1988 on *Shaping Written Knowledge: The Genre and Activity of the Experimental Article in Science,* Miller is cited as identifying genres "as typified rhetorical actions based in recurrent situations" (Bazerman 1988: 159). When genres gain the power to affect society, others wanting to address similar situations must use similar strategies, or genres, to gain a voice. This does not mean that the genre is a static monument in stone. Rather the genre itself continues to be shaped, to evolve, as new texts are created with variations that contribute to its development, usually reflecting changes in the social situation.

This dynamic state of genre is well presented by Berkenkotter and Huckin (1995) as "a sociocognitive theory of genre" which is "grounded" in ample empirical case study research. Their thesis is that "genres are inherently dynamic rhetorical structures that can be manipulated according to the conditions of use, and that genre knowledge is therefore best conceptualized as a form of situated cognition embedded in disciplinary activities" (1995: 3).

1.10 A definition of genre and a proposal of 'bridge' genres

One of the important aims of this book is to develop a way to support the participation of non-native English speakers, who can bring in alternative viewpoints, in a predominantly English-speaking scientific discourse community. With this aim in mind, a heuristic definition of 'genre' is needed. The work of Miller mentioned above offers useful hints toward such a definition: "a rhetorically sound definition of genre must be centered not on the substance or the form of discourse but on the action it is used to accomplish" (Miller 1984: 151). This component of 'social action' is an essential part of Miller's definition. This leads her to reject a class of Environmental Impact Statements that she had studied as not constituting a rhetorical genre because "[it did not achieve a rational fusion of elements – in spite of obvious similarities in form and substance, and in spite of a recurring rhetorical situation that was, in fact, defined by law. These documents had no coherent pragmatic force [...] Overall, the imperfect fusion of scientific, legal, and administrative elements prevented interpretation of the documents as meaningful rhetorical action" (Miller 1984: 164).

Of course, Miller's aim was to define 'genre' from a rhetorical standpoint and differs from what I am trying to do here. However, Miller's statement in itself recognizes the existence of substance and form as well as action; in her case, she places emphasis on the rhetorical function. Here, in trying to develop a heuristic definition, let us turn to the substance and the form. If we consider a genre to be composed of all three, then perhaps we can tease them apart as follows:

(1) Substance – knowledge, information

(2) Form – grammatical, technical, rhetorical (classical sense)

(3) Action – rhetorical (community action sense)

Berkenkotter and Huckin (1995) go a step further and present five features that they claim genres possess: dynamism, situatedness, form and content, duality of structure, and community ownership. Dynamism comes from the change with time and situatedness from the active participation in the "ambient culture". Form and content are both essential to the notion of timely purposefulness. Duality refers to the fact that the structures which genre rules construct are both 'constituted' and 'reproduced' by the people involved. And finally, genre conventions are 'owned' by the community. Their definition is complex because it includes the features of the discourse community itself.

In this book, the definition of 'genre' will be focused on how it can be useful to a non-native speaker user who would need to know the features of a professional genre in order to gain a voice in the discourse community. Only in this way can a non-native speaker contribute to the evolution of genre and fact in science and technology. For the work presented in this volume, 'genre' is defined as follows:

> Genre is a type of recurring communicative event of specific substance, form and action in a focused global discourse community which is recognized by the communicator (writer, in the case of the texts used), who is a member of the community, and others in the community but may or may not be aimed at a receiver (audience, or the reader) in the same community.

Note the removal of the restriction of having the text being limited to the discourse community in which it originated. This is because the idea of 'bridge' genres will be proposed. A discourse community uses 'bridge' genres to interact with other communities. At least two types of bridge genres can be suggested: one which disseminates information from one community to another (e.g. medicine package inserts from a pharmaceutical manufacturer's community to the general public) and another which expresses a current gestalt of one community so that it can be accessed by and perhaps have an effect upon other communities (the review article, popularizations of science articles from a focused discourse community to a discourse community with a wider focus and to a general educated public). The caveat 'at least' reflects the notion that there is very likely to be a 'time' element involved, such as is evident with the botany genres mentioned by Swales (1998) in which the audience for which the text is prepared may not necessarily be present at the time of its preparation. The review article will be shown to be a type of bridge genre.

1.11 A definition of genre with implications for non-native English speaker users

Constant change and evolution, as discussed above, are essential elements in the natural world. In the case of discourse communities and their gestalts, constant innovative input is essential. Such creative contributions can come from different cultures but problems with genre usage can arise if the contributor is, in today's world, a non-native speaker of English. Thus, it is important to find ways to identify the features of a genre so that they can be accessible to non-native English speakers who wish to gain a voice in a discourse community.

Informed by Miller's (1984) analysis, the components of a genre have been identified as being: substance, form and action. The substance component would include the kinds of information that the discourse community would expect to find in that genre. The form component can actually be separated into two levels, one being a 'basic' language level at which the rules common to the grammar of the natural language being used would apply, and the other being close to the 'text type' which Paltridge (1996) points out needs to be distinguished from 'genre.' The former 'basic' level of prescriptive grammar and format rules of, in this case, English, is one of the major obstacles for the non-native English user but, fortunately, can be dealt with using a range of computer-support tools which are rapidly becoming available. As for the latter text-type level, Paltridge (1996) refers to Biber's work (1988) in which 'genre' is considered to be based on external criteria and 'text type' to refer to features of linguistic form, which are independent of the genre. This is a useful distinction because similar linguistic forms can be used in different combinations to achieve different purposes. For example, a problem-solution text type can appear in a research article but also can be found in a range of texts from advertisements to stories (Hoey 1983).

The action component is perhaps the most difficult to define and thus to provide support for. However, this 'problem' is also the key to genre evolution and meaningful contributions to the professional community. This will be discussed in detail below.

In sum, in this volume, texts will be analyzed for features at the levels listed in Table 1. This table suggests that the groups claiming jurisdiction over the microelements of the genre features differ, and this is very likely to be a source of confusion such as that a non-native speaker of English encounters when trying to publish in a global discourse community. By 'claiming jurisdiction' I do not mean to imply that there are strict rules, but rather that the types of patterns have become accepted by the communities using them and thus can be thought of as "highly systematic" (Biber et al. 1999: 5). In fact, because these 'rules' are often not precisely expressed, they are difficult to grasp and follow. However, if these rules are not adhered to, the result can be loss of an opportunity to publish a manuscript and the notion of 'linguistically challenged' members of a global discourse community.

Feature	Realization	Groups claiming jurisdiction
Substance	Information	Discourse community
Form	A. Expression – 'basic' rules of language usage B. Expression – text types	Language community Language community Discourse community
Action	Purpose	Discourse community

Table 1. Genre features and jurisdiction.

1.12 Exploiting genres

Expert members of the discourse community exploit genres "to achieve private intentions within the framework of socially recognized purpose(s)" says Bhatia (1993: 13), adapting from Swales. The notion of 'exploitation' rings of economical advantage. This idea of science for 'profit' was explored by Latour and Woolgar (1982) with work in science viewed as a "cycle of credibility" with the "conversion between one type of capital and another [being] necessary for a scientist to make a move in the scientific field" (1982: 35). Hagstrom (1982: 21) also points to the fact that "manuscripts submitted to scientific periodicals are often called 'contributions'" although he then goes on to distinguish them as "gifts": "gift-giving in exchange for recognition is an appropriate method of social control in science" (1982: 32). Perhaps not as pointedly 'economically' driven, Bex still uses the 'contribution' metaphor: "[...] language is social, but [...] language users are making individual contributions to the societies of which they are a part" (Bex 1996: 3).

What these metaphors point toward is a distribution of 'power' in the community, and Berkenkotter and Huckin have even entitled their 1995 book, *Genre Knowledge in Disciplinary Communication: Cognition/Culture/Power*. We now recognize that in order to participate in disciplinary communication, we must become involved in a 'power struggle' in order to gain a voice. In order to democratically allow participation by cognitively qualified candidates, we need to be able to separate the substance and action from the form of a genre. The form, although an important component, has often been used as a gatekeeping device to shut out what might be valuable substance and action. Of course, the segregation of form is not a simple matter because it is so integrally involved in the creation of the genre. However, this is what will be done in this volume, at least to try to make the genre more accessible to participant candidates whose native language is not that in which the genre we are discussing is embedded.

Miller (1984: 165) ends her paper on "Genre as Social Action" with the statement that "for the student, genres serve as keys to understanding how to participate in the actions of a community". I hope to suggest ways to find these keys.

2. 'Nested' structures, communities and issues

> I argue that words and their relations can be used as an operationalization of epistemic networks in scientific literature as conceptualized in the pragmatic tradition in the philosophy of science, but that one has to add the notion of 'nested structures' of networks as different levels of aggregations [...] in order to achieve a fuller understanding of the organization of texts.
>
> (Leydesdorff 1991: 75)

2.1 Problems of the 'linguistically challenged' scientist

"No language, no facts" (Gross 1990: 203) succinctly expresses the source of problems which can arise in today's world where English is the predominant language of academic communication. Presented here is a case for the status of the non-native English speaking scientist as one of being 'linguistically challenged'.

One of the main aims of this book is to suggest ways to increase diversity and creativity in science by helping to promote contributions to professional discourse communities by scientists who are non-native speakers (NNS) of English. That a problem exists is evident from a perusal of the information for authors/contributors in the journals which warn that "both correct and idiomatic English" (Information for Authors, *Analytical Biochemistry*, 1998) is required for publication. If these conditions are not met, comments such as the following will be returned with a manuscript submitted for consideration for publication:

> Despite the manuscript being reviewed for English grammar, it is almost impossible to interpret. Too many strange sentences that destroy any impact for conclusion. [Comment received by a Japanese scientist from a referee for an American journal, December 16, 1998]

Some journals automatically send manuscripts from 'overseas' authors out to a rewriter after acceptance for publication: "When the Editor has decided to accept a manuscript, it will be sent for editing of the English by a rewriter selected by the Society. The edited manuscript will be returned to the authors for their checking for correctness of the meaning of the revised sentences [...]. Charges for English revision may be made if extensive rewriting is necessary" (Instructions to Overseas Authors, *Bioscience, Biotechnology, and Biochemistry*, 1998). Interestingly, this journal is an English-language journal published by the Japan Society for Bioscience, Biotechnology, and Agrochemistry.

Other journals even recommend editing before submission, e. g.: "Since this *TRANSACTIONS* publishes only in English, contributors for whom English is not a native language should consult a colleague who is familiar with the English language, prior to submitting the paper, for the purpose of editing the original manuscript" (Information for Authors, *IEEE Transactions on Systems, Man, and Cybernetics*, 1998).

Such examples may be instances of what Swales (1998: 207) wonders about: "Will it turn out to be the case that most (or even all) 'true' discourse communities are defensive – that internal coherence is created, reinforced, and ratified by external opposition, disregard, or disdain?" The very discourse that binds can also be used to exclude, as McKenna (1997: 191) points out with a quote from Killingsworth and Gilbertson (1992: 162): "discourse 'is the means by which communities develop and advance their agendas of action, build solidarity, patrol and extend their boundaries, and perpetuate themselves in the life of a general culture.'" In other words, communities use discourse to extend what can be considered their "cognitive territory" while protecting the rights of the privileged members ("build solidarity") allowed to advance the community's agendas.

This 'us-them' viewpoint could contribute to the view of non-native English speakers being considered as 'false outsiders' of a discourse community in which the main medium of communication is English. These people would have some of the credentials needed to join the community because their world view of a discipline can be classified in the same set as that of others in a focused global discourse community. However, because the medium of interaction is language, they are disadvantaged by being unable to fully exploit its forms. The heuristic definition of genre offered in the previous chapter enables us to segregate the different problems which can arise when attempting to communicate via a certain genre. Separating the features of a genre into Substance, Form and Action suggests that the linguistic problems encountered by NNS can be attributed to issues related to the form of a genre. If there are problems with the Substance or Action of a work, then it would not, and should not, be accepted by the community. However, if the problem is only linguistic because of second-language usage issues, then it should be possible to offer help in a practical way to promote more active NNS participation in their discourse community. Of course, because the form of the genre is so intricately involved with its realization, not being able to handle the language form properly is often mistaken as not having something of value to contribute.

This complex interaction arises from the fact that the genre both constitutes the structure and is generated through community interaction. An example of how this occurs is described by Berkenkotter and Huckin (1995) who refer to work in organizational communication by Yates and Orlikowski (1992). They traced the evolution of an institutional genre, the office memo, and showed how it was affected by the advent of the typewriter and then by the formation of the job of typists which then led to standardization of document format. Because of the rules of document format which arose from this type of institutional genre, a person wanting to contribute to a professional journal will find very minute details about format specifications which need to be cleared in order to have a paper published. This is one of the levels of difficulties facing a non-native speaker, or even a native speaker of English who is a novice to the community. The latter can be expected to have an advantage of possessing an internalized linguistic system that serves as the basis of the professional discourse.

2.2 The notion of discourse community 'jurisdiction'

To examine the problems more closely, let us consider it from the definition of genre used in this book (section 1.10). As discussed in Chapter 1, the features of a genre were separated into Substance, Form and Action. Problems in having work published can arise at different levels because different communities have 'jurisdiction' over different portions of a genre. The multiplicity of discourse communities was introduced in Chapter 1, with the idea that an individual can, and most likely does, claim concurrent membership in several discourse communities. The notion of 'jurisdiction' does not mean that there are explicit rules but rather that communities, by consensus, appear to have notions about what is acceptable and what is not. In fact, the very lack of explicit acknowledgement of 'jurisdiction' is likely to be one of the obstacles to NNS participation in an international discourse community. If there were explicit rules, it would be a simple matter to conform to them. However, if these "rules" and a definite form of the 'community' using them are not explicit, then it is difficult to grasp what is required. Although not explicit, these 'rules' or expectations nonetheless exist. Perhaps this can help explain the statements presented in the Instructions to Authors of the journal quoted above, which are some of the simpler rules that need to be observed. While they may be easier to observe than, for example, rhetorical argument expectations, they remain as stumbling blocks which can jeopardize community participation.

As a specific example of the 'inexplicit' yet 'real' rules and expectations that pervade the academic discourse communities being considered here, let us turn to an interesting report from Berkenkotter and Huckin (1995: 97-116), who devote a chapter in their book on *Genre Knowledge in Disciplinary Communication: Cognition/Culture/Power* to "Gatekeeping at an academic convention". They studied the proposal reviewing process of the 1993 Conference on College Composition and Communication (CCCC), "the primary professional organization of postsecondary writing instructors in the United States" (Berkenkotter and Huckin 1995: 97). They note that "submitters are given relatively little guidance in how to write an effective abstract" and that the members themselves showed "widespread uncertainty about what constitutes a good CCCC abstract" (Berkenkotter and Huckin 1995: 97-98). Despite this noncommittal official attitude, in a pilot study, they discovered two types of features, general and distinguishing, in the favorably rated abstracts. In the latter, they found that the distinguishing features could be further categorized into abstractness, addressing of social issues, and more complex and problematized. To further identify these features at the text level, they conducted counts of indicators of these features and found a statistically significant difference between the higher and lower ratings. Thus, despite the fact that no definite instructions had been given to the proposal writers or the evaluators, the discourse community seems to have possessed an awareness of what was acceptable and what was not. It is this type of 'jurisdiction' that I am trying to address here. The fact that it is not overtly expressed makes it all the more difficult to identify and deal with when problems arise.

Belcher and Braine (1995: xv) also point to a need for "demystification of Anglophone academic discourse" because of "a belief in the power of explicit cognitive awareness of the texts, subtexts, and contexts of academic discourse to enable individuals to join the collectivist endeavors that academic communities are [...]" engaged in. They call for a "shared awareness of the rules of academic games and the strategies that successful players use" and while recognizing that they cannot teach all of the required 'passwords,' can try to "teach some subset of them, and through these, a feeling for how they function rhetorically, the intentions they can serve, and the messages readers can construct from them." Here also is the notion of the 'passwords' of academic discourse not being explicitly expressed but rather present as a shared awareness by the community members. Belcher and Braine (1995) devote an entire book to "demystifying" these "passwords."

2.3 'Nested' structures

When considering the second-language issues involved in publishing in research science journals, what we seem to have is a set of 'nested structures' which complicate matters because they mutually interact with each other. The term 'nested structures' is borrowed from Leydesdorff (1991) who is working in 'science dynamics' which uses scientometric methods for science studies. Although lacking a discourse community dimension, his work can be useful when examining the text itself. He uses the work of Mary Hesse (1980), following Quine, to posit 'epistemic networks' which are defined "as structures of science which are retrievable in terms co-occurrence and co-absence in scientific texts" (Leydesdorff 1991: 77). He then goes on to state:

> the proponents of the sociology of translation have introduced the concept of the sciences as 'nested networks': [...] analogous, for example, to the way in which actors may behave on behalf of groups in constituting the state apparatus. However, as a methodology in this tradition, the various levels of aggregation are juxtaposed: each actor-network may be composed of other actor-networks which may interact, and be compared with other actor-networks at other levels of aggregation. (Leydesdorff 1991: 79)

My interpretation of this is that there are many levels of interaction involved in the publishing of research in written form – scientists doing the research (usually a group rather than an individual), their colleagues in the field, the journal editors and reviewers. All of these 'actors' are members of their own discourse communities [e. g., there is a community of biology editors who have published a *CBE* (Council of Biology Editors) *Style Manual*, 5th ed. (1983)] and therefore would have their own expectations to fulfill. The resulting interactions among these people would be complex and occur at various levels.

This notion of 'nested' structures can perhaps help illuminate the problem pointed out by Paltridge (1996: 237): "the terms 'genre' and 'text type' seem to have been conflated, with the term 'genre' being used to include both of these notions". He illustrates the differences between the two with a formal business letter and the abstract from an experimental research report, using the same texts to point out both the generic structures and the text structures. I would like to suggest that the genres and text types are sometimes mistakenly mixed because the elements of both are present in the same text and interact with each other. As stated in Chapter 1, the form portion of the genre concept used in this volume includes the text structure of Paltridge while the action portion would be equivalent to his generic structure. Although Paltridge has neatly distinguished between the two structures, the distinction may not be as clear when the entire text is considered against the expectations of the discourse community.

2.4 'Nested' communities

2.4.1 English-speaking community (ESC)

Here let us extend the term 'nested' to signify the juxtaposition and interaction among the communities involved in the genre we are going to examine in this book – the review article. The problems of non-native English speakers are complicated by the fact that they are trying to participate in at least three different communities at the same time: the English-speaking community, the English-speaking academic community, and the focused global discourse community. At a basic level is the English-speaking community which uses English as the main language of communication and therefore has developed its own rules and regulations concerning the usage of the language. The effects of this community are mostly at the 'lowest' level of basic grammar rules; word choice, especially pertaining to phrasal verbs, idiomatic expressions and other collocational issues; technical details, such as punctuation and spelling. The *Longman Grammar of Spoken and Written English* (Biber et al. 1999) attests to the existence of such a community. It is important here to note that this 'community' is not being referred to as a 'discourse' community. The ESC is a group of people who recognize different registers in language usage but are too loosely bound (mainly by the language in general rather than by specific genres) and do not have meaningful focused social interactions such as observed in 'discourse' communities. However, as 'native' English speakers, their awareness of language expectations is at the basis of the expectations of the other two communities: the English-speaking academic community and the focused global discourse community.

2.4.2 English-speaking academic community (ESAC)

At another level, there is the English-speaking academic community composed of members who are aware of what communications will be considered acceptable within it. This community is also rather loosely defined though in a more restricted sense than the English-speaking community. In addition to the work of Belcher and Braine (1995) introduced above, the expectations of this community are clearly presented by Johns (1997: 58-64) who summarizes the work of Elbow (1991), Geertz (1988) and Purves (1990). Although the list is a bit long, it so well captures the 'Ten Commandments' of the English-speaking academic community today that it is presented below. Another reason for including this list is that reference to it will be valuable in identifying form constraints determined by the English-speaking academic community. From Johns' (1997: 58-64) book *Text, Role and Context*, here is a list on "the nature, values, and practices in general expository academic prose":

(a) Texts must be explicit.
(b) Topic and argument should be prerevealed in the introduction.
(c) Writers should provide "maps" or "signposts" for the readers throughout the texts, telling the readers where they have been in the text and where they are going.
(d) The language of texts should create a distance between the writer and the text to give the appearance of objectivity.
(e) Texts should maintain a "rubber-gloved" quality of voice and register. ("They must show a kind of reluctance to touch one's meanings with one's naked fingers.")
(f) Writers should take a guarded stance, especially when presenting argumentation and results.
(g) Texts should display a vision of reality shared by members of the particular discourse community to which the text is addressed (or the particular faculty member who made the assignment).

(h) Academic texts should display a set of social and authority relations; they should show the writer's understanding of the roles they play within the text or context.
(i) Academic texts should acknowledge the complex and important nature of intertextuality, the exploitation of other texts without resorting to plagiarism.
(j) Texts should comply with the genre requirements of the community or classroom.

2.4.3 Focused global discourse community (FGDC)

Finally, we have a third-level community which is the one the would-be participant is most aware of, the focused global discourse community. The problems which arise from the expectations of the other two communities are often mixed with those of this immediately evident community and therefore complicate matters. Although a clear separation of the issues would probably not be possible, even an awareness of this interaction among communities should help identify ways in which to cope with problems which can arise in academic genre writing.

2.5 'Nested' issues

The above discussion suggests that the problems with English usage arise at several different levels for non-native speakers of English (NNS). At the risk of seeming to be oversimplifying the issues involved, these problems are presented in Table 2. Through my work with Japanese scientists trying to publish their manuscripts in English, I know that it is often not possible to clearly state why an expression needs to be changed or clarified. However, one of the main aims of this work is to propose a heuristic tool for NNS. In order to do this, some sacrifices may be needed. This is an approach used in engineering. For example, in trying to develop a robot that could walk on two legs, the characteristics of walking had to be simplified to the point that the first upright 'walking' robot was actually a one-legged 'hopping' robot. Only after the problems of upright balance were solved using this prototype could the research and development proceed to a two-legged walking robot.

In the case of multifacted linguistic issues, some over-simplification is unavoidable in initial steps to produce a useful heuristic tool which can then become the basis of further development. I have therefore categorized the problems according to type in order to discuss how to find ways to deal with each type of problem, with some involving 'jurisdiction' at more than one level. Only the basic ideas of the nested issues are presented in the following sections because they will be discussed in more detail with examples later, which proposes ways in which help can be offered in resolving these issues for NNS writers.

Related genre feature		Community claiming jurisdiction	Type
Grammaticality	Form	ESC	I
Technical format	Form	ESC, ESAC, FGDC	I, II
Sentence level expectations	Form, Substance	ESC, ESAC	II
Metatext features	Form	ESAC	III
Paragraph level expectations	Form, Substance	ESAC, FGDC	III
Rhetorical organization	Form, Action	ESAC, FGDC	III

Table 2. Community jurisdiction in problems with genre features.
(ESC = English-speaking community;
ESAC = English-speaking academic community;
FGDC = Focused global discourse community)

2.5.1 *Type I issues*

Grammaticality and technical format are among the first hurdles as is evident from the instructions that science journals give to authors. From the viewpoint of a non-native speaker, these are the features usually covered by English grammar books and textbooks – grammar and technical details including spelling. Today, support for many of these features, particularly the ESC ones, can even come from word-processing programs which can highlight possible misspellings and grammatical problems. Of course, these tools are far from adequate but can serve as a step in the direction of demystifying professional discourse. If more commonly used conventions can be discovered, more appropriate technical support may be possible. For example, the advice to use the active rather than the passive voice is often overgeneralized and therefore not helpful for some types of academic writing. In fact, the use of the passive might even be suggested for certain types of texts, such as the experiment procedural explanations in a science report.

Perhaps these technical details seem very simplistic but for a nonnative speaker trying to compose a 'genre' text for a professional discourse community audience under time constraints, help with the basics can free the writer to deal with the higher level content issues. As an example, I myself as a second language learner of Japanese can compose texts belonging to a certain genre in Japanese only with the aid of a Japanese word processor program that will handle the kanji characters in the complex Japanese writing system. Without this kind of basic aid, it would be extremely difficult to lead an active professional life in Japan.

Issues related to technical format also involve the ESAC and FGDC because there are many specific conventions, such as for measurement units and symbols in the sciences, which require specific knowledge to address. Below is an example of a checklist enclosed with a letter from the managing editor (June 25, 1991) to the author of a paper submitted to the American Heart Association journal *Stroke* (Observe the 'power' overtones in the final sentence):

> Notice to Authors: Your manuscript does not conform to all the requirements of the STROKE format as published in the January and July "Guidelines to Authors." Please note the following and revise your manuscript accordingly. Failure to do so will delay consideration of your article.

The notice was a list of twelve items, of which five had been marked with arrows for particular attention. Here are some of them dealing with the use of abbreviations and with paragraph and format conventions:

- Abstracts of original contributions should not exceed 200 words and must contain no abbreviations. This summary should state the purpose of the study; the materials and methods employed, including statistical significance; the results obtained; and a brief conclusion or statement of potential relevance of the work. [Underlining handwritten]
- Avoid the 'outline' format by deleting all paragraph headings other than the customary ones for original contributions: Introduction, Material and Methods, Results, Discussion, and Acknowledgments (if present). Each section should begin on a separate page.
- In general, keep abbreviations to a minimum throughout the article, using only those most commonly recognized.
- *Spell out FDP, BA, SCA, PCA, DAT, PAP, DM, HL, HT, AF* [Handwritten]
- There are also spelling conventions which must be observed.
- Names of chemical or organic substances should follow the recommendations of the IUPAC-IUBMB Joint Combined Commission on Biochemical Nomenclature as published in the extended Instructions for Authors [...] (*Analytical Biochemistry* 1998)

These Type I issues are generally clearly defined and usually are not related to the content of the text itself. Therefore, these 'surface' conventions can usually be easily followed. However, if these conventions are not followed, the result can be a delay in publication.

2.5.2 Type II issues

Type II problems arise mainly, but not only, from the expectations of the English-speaking academic community. Also, these issues are not restricted to the academic community. They are more complex as they are, in part, influenced by cross-linguistic issues. The Type II issues can generally be separated into those involving word choice and those involving sentence level structures. The word choice issues will be focused upon later when dealing with lexicogrammatical support tools and will therefore not be discussed in detail here.

The other portion of Type II issues is the sentence structure problems which can arise because of differing grammatical conventions between the first language and English. To illustrate such a problem, here is an example of a sentence-level information-ordering problem often encountered in texts written by Japanese-language speakers:

(1) So far, allelic loss, gene mutation, and DNA methylation in promoter lesion are reported as possible mechanisms for X reduction.

Revision suggestion:
As possible mechanisms for X reduction, allelic loss, gene mutation, and DNA methlyation in promoter lesions have been reported.

Here are some other examples of sentence-level problems commonly encountered by Japanese writers.

(2) Thus, for colorectal adenocarcinoma, it is more useful to investigate the expression of X as well as that of Y for predicting tumor invasion and metastasis than examining Y only.

Revised version:
Thus, to predict tumor invasion and metastasis in colorectal adenocarcinoma, not only expression of Y but also that of X needs to be examined.

Comment: The aim of the study can be more quickly grasped if the phrase dealing with the purpose comes earlier in the sentence.
The original sentence came from a paper which had submitted to an American journal but was returned because "both reviewers felt that extensive linguistic editing by an expert in English would be necessary" despite the fact that the content was considered acceptable for publication.

(3) [...] However, the number of markers is still insufficient. From this standpoint, the present contig must be reexamined using a larger number of landmarks.
Recently, RG was developed as a method to scan a large number of restriction sites distributed on entire genome. RG employs [...]

Revised version:
[...] However, the number of markers is still insufficient. From this standpoint, the present contig must be reexamined using a larger number of landmarks.
One solution to this problem is offered by RG, a method developed to scan a large number of restriction sites distributed on an entire genome. RG employs [...]

Comment: Adding the discourse signal 'one solution' [...] tells the reader what to expect.

(4) The influence of the kind of alkali (KOH, NH_4OH, ethanolamine) in alkali treatment, on the formation of multi-hollow structure within submicron-sized styrene-methacrylic acid copolymer particles, by the alkali/cooling method proposed by the authors, were examined.

Revised version:
We examined the influence of the kind of alkali (KOH, NH_4OH, ethanolamine) in alkali treatment on the formation of multi-hollow structure within submicron-sized styrene-methacrylic acid copolymer particles by the alkali/cooling method which we proposed earlier.

Comment: The agent and the verb play important roles in moves and should be presented as soon as possible to enable the reader to judge the rhetorical move of the sentence.

To deal with these Type II issues, the expectations of the discourse communities involved need to be clarified. The micro-level analyses will be presented in later chapters to suggest ways in which these issues can be resolved.

2.5.3 Type III issues

Classified as Type III are paragraph and text issues of information presentation and ordering, including the need to consider cross-cultural issues. Here Johns' 'Ten Commandments' (1997) can be very useful for identifying ESAC expectations, e.g. "Topic and argument should be prerevealed in the introduction" (Johns 1997: 58). Cross-cultural issues are involved because, for example, at the paragraph level, the Japanese sequence often proceeds from specific to general while English, especially in academic writing, prefers the organization pattern of general explanation with supporting details. This is discussed in more detail below.

Type III problems are not only those of information ordering but also of whether or not the information is included. For example, Mauranen, in the conclusion of her textlinguistic study on academic writing in Finnish and English, states (Mauranen 1993a: 253):

> Text about text has a guiding, organising and a clarifying function and it reflects a writer's awareness of the text as text. A rhetorical strategy which is highly reflexive can be characterised as personal, explicit, and helpful. It guides the reader, making him or her aware of the organisation, functional parts, and central messages of the text. This appears to be at least a general Anglo-American attitude.

Type III problems often arise with Japanese writers. Some editorial comments received by them are shown below.

(5) This study is interesting but should present more clearly the purpose of the work, is there an hypothesis for testing, and of what significance are the results. An important consideration is improvement of the writing of this work. Editorial assistance of one familiar with English and this work would be helpful. [anonymous referee of *Cryobiology* commenting on a paper submitted for publication by a group of Japanese scientists, 24th November 1995]

(6) Your manuscript entitled [...] with your various colleagues has been reviewed by the Editorial Board. I am pleased to inform you that they find the article interesting and important and recommend that it be accepted following your attention to certain concerns. In the main, the problems raised relate to the presentation of the results. They include your use of the English language, and the materials and methods section which is too brief and confusing figure legends. Hence, all of their concerns have to do with manuscript construction rather than technical concerns. Nonetheless, these are important points and must be attended to before the article can be accepted in *The Journal*. I recommend that you seek the assistance of an individual for whom English is their native tongue to correct the problems with usage and grammar and to aid you in rewording the figure legends. [Associate Editor, The Journal of Biological Chemistry, American Society for Biochemistry and Molecular Biology, December 2, 1994]

One reviewer specifically noted that there were problems with the English throughout but a glance at the problematic sentences he quotes would puzzle even the native English speaker as to what was wrong with their 'English':

(7) The clinicopathologic importance of the biologic aggressiveness has been well documented in many reports.
The classification by Sakamoto et al defined both papillary and follicular carcinomas as poorly differentiated carcinomas.

Yet the reviewer specifically states the following which indicates the importance of the current viewpoints of the FGDC in text considerations:

(8) There is a problem with the English throughout the text. It is not a very serious one, but it certainly detracts from the message and makes some important statements not immediately intelligible. Among the many examples I could quote, I will select these:

'The clinicopathologic importance of the biologic aggressiveness has been well documented in many reports.' (First sentence of Discussion, page 8). What does this sentence mean? I think the authors are trying to say that some clinical and pathologic parameters of thyroid carcinomas have been found to correlate with the tumor aggressiveness, but it sure takes a while to decipher the message.

'The classification by Sakamoto et al defined both papillary and follicular carcinomas as poorly differentiated carcinomas.' I assume they are trying to say that Sakamoto's poorly differentiated carcinomas include tumors in both the papillary and the follicular category. [Anonymous reviewer for The American Journal of Surgical Pathology, December 1997]

While there may not be clear definitions of the 'rules' that are being transgressed, the fact that reviewers and editors point to the existence of problems indicates that there must be some criteria not being met. Perhaps they cannot be labeled as rules but they are real obstacles to the NNS. If the present book were simply one aimed at informing the NNS scientist about how to cope with these issues, the next step would be to offer possible solutions. However, as discussed in Chapter 1, the basic position taken here is that science is a worldview which uses language as one means of expression and that there are multiple possible worldviews. Therefore, rather than simply present 'how-to' hints, let us probe for ways in which alternative worldviews can be expressed. In order to do this, we must first establish that there are alternative ways of expression, or alternative rhetorics. The remaining portion of this chapter will explore this notion using Japanese as an example language together with a consideration of genre characteristics.

2.6 The notion of a frozen genre

The discussion in this chapter has been mainly concerned with community expectations related to academic writing and how to deal with them. Certainly, the simplest way to have work published is by conforming to what is expected. However, as discussed in Chapter 1, discourse communities are dynamic and in constant flux. Each current state is but a metastable one, or a temporary but relatively long-lasting non-equilibrium state. Without this metastability, there would be no progress nor would there be 'life'. One of the absolute requirements of a dynamic entity is constant action. Once the action stops, the entity 'dies' or enters a different state of being. A discourse community that stops evolving loses its 'active' genres although it can retain 'frozen' genres, for example, myths that are passed down through oral tradition with little allowance for variation in form.

In the scientific discourse community, the research article may be on its way to becoming a 'frozen' genre. As reviewed in Chapter 1, it has been under development for about the past three hundred years (Gross 1990). It has been scrutinized by many (Swales 1981, Myers 1985, 1989, 1991, Hopkins and Dudley-Evans 1988, Hunston 1989, Gunawardena 1989, Gosden 1992a, Hozayen 1994 and many others). Evidence for its becoming frozen comes from its being described by numerous 'how to write' manuals and books, with many specifically aimed at the nonnative speaker of English: Michaelson (1982), Barnes (1982), Day (1988), Weissberg and Buker (1990), Huckin and Olsen (1991), Swales and Feak (1994). Some examples of style manuals are those published by Council of Biology Editors (1983) and the American Society for Microbiology (1991). All of this is in addition to the rather specific instructions given by some of the journals themselves about what they expect for a research article. When this much is said about one genre, there is a tendency for less acceptance of deviation by the journal editors and reviewers, in other words, it is becoming frozen. The form begins to assume a greater relative importance in comparison with substance and action. We begin to see what Humphries (1998) laments about – scientists as producing 'disembodied' writing.

Of course, this does not mean that the genre of the experimental research article has lost its utility. It is and will probably remain an important genre for reporting the results of empirical science. However, as a genre, it seems to be gradually becoming 'frozen' in form.

2.7 Consideration of alternative rhetorics

If a science discourse community promotes a view of science reached via an ongoing discourse exchange among its members, then further development of the community will also occur via discoursal input. However, if the genres of a community become frozen, then although the total cognitive quantity of the community may continue to increase, qualitative development may not readily occur. This can have a serious effect on the viability of the community as a whole because dynamism and metastablity are essential features of an evolving 'living' entity. In today's globally interconnected world, this dynamism could arise from the acceptance of more alternative rhetorics in the world of science. As discussed above, because of the intricate entwining of rhetoric and cognitive concepts of science, this is more easily said than done. However, there is a science genre which may be more tolerant of the introduction of alternative rhetorical styles – the review article.

The review article will be discussed in detail in subsequent chapters. Briefly, it appears to be a genre under development with few restrictions as to format despite the fact that it is recognized by the discourse community as a genre, as is evident from written instructions in journals and interviews with specialist informants in the community, which are presented in Chapter 3.

Alternative rhetorics have been the subject of study since the attention-getting 'doodles' paper by Kaplan in 1966 suggesting the existence of different rhetorical patterns in different languages based on observations of the writing of students of English as a second language. Many other reports followed on the topic (Hinds 1983, Connor 1987, 1996) and studies abound in cross-cultural rhetoric or intercultural communication (see Connor (1996) for a recent summary and synthesis). Kaplan was criticized for his simplistic conclusions and 'revisited' his work (Kaplan 1987, 1988), but as Connor, herself a Finn who encountered difficulties as a graduate student of English literature and linguistics, points out in her interview for *The Language Teacher* (published by The Japanese Association for Language Teaching), "the fact remains that EFL learners experience problems in writing in English and many of these problems seem to be ones of transfer.

So, contrastive rhetoric, an area of research in L2 acquisition which identifies problems in writing by referring to the features of L1, is important for teachers and teacher educators" (Muramatsu 1997: 7).

The extreme visually expressed difference between a linear (English) pattern and a circular (Oriental) pattern presented in the 1966 paper by Kaplan attracted much attention to contrastive rhetoric, especially on Oriental patterns. Although a difference does seem to exist, the matter is of course not as simplistic as the doodles make it appear to be. Also, these differences are not only between 'the East' and 'the West'. The writing by Finn academics was the focus of research by Mauranen (1993a). In comparative studies of English and Finnish texts, she examined the use of references, theme, text reflexivity and text structures by Finns and Anglo-Americans and found that the writers used "culture-specific rhetorical strategies, while conforming to the common norms of the genre" (Mauranen 1993a: 263). The Finns used fewer text references and less information in them, had problems with L2 English themes, wrote less metatext, and delayed presentation of the main point. This led her to call for teaching textual phenomena in the same way as lexicogrammatical features are taught. As for the reason for this, she states (Mauranen 1993a: 263):

> The option of not conforming to the norms of the target linguistic culture is not available with respect to grammatical and lexical use, and, as it seems, at least some textual rules must be included in the same category, possibly more than we are accustomed to thinking at present. Breaking grammatical rules has different consequences from breaking textual or rhetorical rules originating in a national culture: by breaking grammatical and lexical rules, a writer conveys the impression of not knowing the language, which may in mild cases be forgiven and in serious cases cause breakdown of comprehension; by breaking rules of a textlinguistic type, a writer may appear incoherent or illogical; finally, by breaking culture-specific rhetorical rules a writer may seem exotic and command low credibility.

These observations may help explain why 31 percent of all scientific papers published in mainstream journals are from the United States, according to Gibbs (1995) in an article aptly entitled "Lost science in the Third World". Japan accounted for 8 percent but when it came to Brazil, Mexico and Egypt, the percentages plummeted to 0.6 to 0.3 percent. Swales refers to Bailey who reports that a Pakistani commentator once said: "Teaching English has become a multi-million-dollar business the world over, a lucrative business next only to drug trafficking" (Bailey 1991, in Swales 1997). The message might be summarized as "publish in (acceptable) English or perish".

2.8 Focus on Japanese rhetoric

Rather than attempt a broad-ranging comparison of various languages from literature data, the focus will directed at the Japanese writer as my experience has been mainly in the Japanese milieu. An in-depth examination of one language should offer insights into how contributions might be elicited from other languages.

To examine the 'circular' pattern of Oriental languages, Hinds examined Japanese (1986, 1987) and pointed out the use of the *ki-shoo-ten-ketsu* pattern in Japanese writing which can be traced back to the Chinese classics. This pattern is also used in Korea (*ki-sung-chon-kyul*) according to Eggington (1987) whom Connor (1996: 45) cites as having done innovative "rigorous text linguistic analysis of Korean texts." Both patterns present "an introduction, the development of a topic, a turning to a somewhat unrelated topic, and a conclusion" found in the Chinese *qi-cheng-jun-he* style (Connor 1996: 45). The three patterns are expressed by the same Chinese kanji characters but the readings differ with the vernacular.

This classical style, however, is recently coming under fire in Japan, as noted by Maynard (1996: 417-418):

> The *ki-shoo-ten-ketsu* tradition which places the conclusion at the very end of the discourse has especially become the target of criticism from Japanese scholars and specialists. For example, Shibata (1992) contends that the imperative of our contemporary "Information Age" necessitates placing the conclusion at the beginning of the discourse. It is strongly recommended that, especially in practical discourse such as that of business documents, mass communication and academic theses, the point should be made clear at the vary (sic) start of the text. Lamenting that Japanese writings tend to be essay-like and often meander to no end, Shibata recommends a reordering of the original *ki-shoo-ten-ketsu* so that some writings follow a *ketsu-ki-shoo-ten-ketsu* order.

Interestingly, Naoumi (1996) found that when Japanese newspaper editorials were translated into English, a synopsis of the editorial's main message preceded the translation – which would be a realization of the ordering proposed by Shibata (1992).

As an example of *ki-shoo-ten-ketsu* writing, Hinds (1987) used the *Vox Dei, Vox Populi* column which appears daily in the *Asahi Shimbun*. This type of column can be found in all of the major Japanese daily newspapers and presented here is an example from the *Jottings* column of *The Daily Yomiuri*, which is an English-language daily published by *The Yomiuri Shimbun* (another major Japanese daily newspaper). The column is a translation of *Henshu Techo* which appears in the vernacular, and the English translation version for May 1, 1999 had originally appeared in the April 29th issue. It consisted of 262 words and 10 paragraphs (the paragraph formatting used in the translated version; the Japanese original usually does not have paragraphs but symbols, such as a diamond or an arrowhead, to indicate breaks in the discourse). There is no title other than the column title *Jottings*. Paragraph 1 is a single sentence: "The recently opened Showakan museum in Kudan, Tokyo, is dedicated to preserving and passing on to future generations the memory of hardships suffered by ordinary people during and after World War II". As there is no title, an English-speaking reader, after reading this first sentence, might expect that today's column will be a presentation of a new museum or the suffering experienced during World War II. As if to verify this, Paragraphs 2 to 5 continue to describe the exhibits and what the author felt when viewing them ("memories of those difficult days came flooding back"). However, Paragraph 6 begins abruptly with "Thursday marks Greenery Day, which used to be celebrated as Emperor Showa's birthday". (April 29th this year was a Thursday.) Paragraphs 7 and 8 report that "some people are calling for this national holiday to become a day to remember the Showa era [...]" Paragraph 9 says that perhaps 'Greenery Day' should be moved to May 4th. The final paragraph ends with a combination of reminiscing about the war and 'greenery': "I remember how badly this land was ravaged during and after the war. And so I would like to thank our forebears for making this country green again". The reader, from an English-speaking viewpoint, might be left to wonder what the entire piece was about – a new museum, reminiscence about World War II, a proposal to change a holiday to commemorate something else, appreciation of greenery. The fact that this type of column appears in every major Japanese daily and is considered 'good' reading suggests that the Japanese public accept and appreciate this type of discourse.

Dennett (1988: 119) explored the use of the *ki-shoo-ten-ketsu* structure in technical writing. She points out that there is a need to "derail any desire to surprise, entertain, or please the audience, instead of informing the audience according to the logical patterns they expect."

This indirect type of discourse may be changing due to the needs of modern society. Japanese television newscasters are often heard to say *"Ketsuron kara moshimasu to"*, which literally means, "I will start (honorific form) with the conclusion [...]". Nagano (1983), who examined news commentary segments from Japanese television programs, reported that the central message is announced at the beginning. What is of interest here is the fact that there seems to be a need to 'announce' a general-to-specific ordering of information, suggesting that it is considered to be a marked structure.

The tendency to use a specific-to-general style is also evident in Japanese student writing. Kobayashi (1984) examined the rhetorical patterns in English and Japanese in 676 compositions written by 226 American and Japanese college students studying in both the United States and Japan. Her study showed that the students "consistently differed from each other in their use of rhetorical patterns and kinds of general statements" with the Japanese students preferring the specific-to-general pattern in contrast to the general-to-specific pattern favored by the American group.

Of course, calling too much attention to the 'uniqueness' of Japanese written discourse can be problematic, as pointed out by Kubota (1997). She warns that there has been too much overgeneralization about the cultural characteristics of Japanese written discourse (with excessive focus on *ki-shoo-ten-ketsu*) based on a limited number and type of examples – the *Tensei Jingo* column originally studied by Hinds (1987) is only one type of newspaper essay and its characteristics should not be generalized for the Japanese language as a whole.

Kubota also points out that Japanese written discourse started undergoing vast changes since the Meiji period (1968-1912) when Japan opened its doors to the West after more than 200 years of self-imposed isolation. Thus, written discourse today has been strongly influenced by European languages, particularly English. She quotes Ishiwata (1971) as stating that around 1878, 70 to 80% of the books published in Japan were translations from Western languages which were prepared with attempts to keep the original structures intact. This influence from other languages has continued and there are even 'how-to' books on composition for Japanese academics which are similar to English composition manuals – one by Kinoshita (1981) is recommended reading for many engineering students. It should be interesting to examine how this flood of Western rhetoric has influenced the Japanese language.

This intricate mix of language and culture was the topic of a stimulating symposium held in 1996 at Osaka University in Japan. Its title was "Linguisticulture: Where do we go from here?" and the keynote speaker Fujimoto states in the proceedings that they coined this word to represent "the birth of a new discipline that sees language and culture as a compound whole, and that we are not here simply here to study a variety of languages and cultures" (Fujimoto 1996: 91). He describes how this idea came about as Japan struggled to find how to achieve "an organic comprehension of a foreign culture where a culture and its language are not separated". With the establishment of a faculty of language and culture in 1974 and then a graduate school in 1989, they were driven to seriously consider the concepts behind this 'linguisticulture' idea. Today, they interpret it as "a complex and interdisciplinary field of research, having contact with all the other disciplines of the natural sciences, social sciences, and humanities".

2.8.1 Alternative linguisticultures – a consideration of Japanese

With this concept of linguisticulture in mind, I would like to present a 'startling' quote from a respected Japanese philosopher, Nakamura, who claims that Japanese is an 'irrational' language. Nakamura tries to delve into the ways of thinking of Eastern peoples to find the roots of 'Japaneseness' or 'Chineseness' or 'Koreaness' via scrutiny of their ways of thinking which can be viewed through, among other phenomena, their language. Here he is discussing Eastern logic and traces its path from India through Tibet and China to Japan. Of interest to us here is his statement that although logic should be considered as "the most universal form of learning," it was affected by the culture with which it came into contact. This occurred because the language used to transmit it varied from people to people (Nakamura 1964: 8-9):

> [...] logic is one of the most important keys for the study of the features of the ways of thinking of a people. Accordingly, by studying Eastern logical works and at the same time comparing them with those of the West, we should be able to comprehend some features of the conscious ways of thinking of the various Eastern peoples. Logic in the East originally appeared in India, but when it was introduced into Tibet, China, and Japan, it was studied in different ways in each place, and in each country it was considerably modified. Although logic should be the most universal form of learning, as a matter of historical fact it has by no means been formulated or transmitted to these other peoples in a universal language. Naturally, the characteristic features of ways of thinking, differing with each people, are reflected in the mutual differences of native or imported patterns of logic when the latter are tied to the structure of a given language. [...] In spite of the fact that the masses use language constantly every day, their use of logical forms of expression is almost non-existent. Consequently, it is incorrect to say that logic regulates the ways of thinking of a given people to the same extent as do linguistic forms.

Before proceeding further, a definition of 'logic' for the purposes of this discussion might be relevant here. According to *The Linguistics Encyclopedia* (Malmkjaer 1991: 129), "logic studies the structure of arguments, and is primarily concerned with testing arguments for correctness or validity". McCawley (1993: 1-2) states that

> logic is of necessity also concerned with semantic analysis, that is, with determining what propositions are expressed by or involved in the sentences of natural languages such as English or Japanese or Wolof: since inferences framed in ordinary language form a major part of the 'data' that logicians must provide an account of, semantic analysis will be a major part of the practising logician's activity and of what must be learned in order to understand logic [...]

If logic in the case of natural languages is intimately involved with its linguistic expression, then the extent to which the thought sequence is made apparent in the surface forms could differ with the language. Some languages would need to make more of the thought process evident than others. In the case of Japanese, there seems to be a greater tendency to assume understanding of the thought processes which lead to an end result. Therefore, attempting to study Japanese from the viewpoint of an outsider not aware of the linguisticulture in the background may lead to the conclusion that Japanese lacks rationality as a language.

The 'irrational' label given by Nakamura to the Japanese language can be traced to his claim that the Japanese people display a marked tendency to "give special attention to those subjective and social relations and actions which form the basis of mutual understanding and loyalties to the family, clan, and nation" (Nakamura 1964: 531). He goes on to point out the fact that Japanese sentences themselves tend to emphasize emotive factors rather than cognitive ones. Thus, this irrationality does not imply that the Japanese have no ability for logical thinking, but rather suggests that the 'logic' traditionally acceptable to Japanese speakers may take forms differing from those acceptable to speakers of other languages, in the present discussion, English speakers. Because of a differing focus of the functions of language, preciseness and accuracy in expression are substituted by vague and typological expressions.

2.8.2 Vagueness in Japanese

Let us now turn to a consideration of the vagueness present in the Japanese language. At the syntactical level, Japanese sentences are very often without a subject and there are no signals for plurality or gender. The absence of the subject can be a cause of vagueness or, from the Japanese point of view, offer a situation that is purposefully rich with possible interpretations. As for the lack of plurality signals, I remember doing translation work for a company that produces lighting facilities and having trouble with the descriptions which offered no precise information on whether I was describing a room with one chandelier or two or more. Of course, Japanese allows the description of plurality but including such information in this type of text would have been awkward, suggesting that the need to know whether there was only one or more than one chandelier was irrelevant to the Japanese reader. Another feature leading to vagueness is the lack of gender reference, which makes it difficult to know whether the person you are to meet is male or female (albeit, in some ways, this can be very convenient in a gender-sensitive linguistic atmosphere). The list of examples can go on when it comes to 'vagueness' in Japanese.

These features suggest that there seems to be a higher tolerance level of vagueness in Japanese that encompasses even the extent of how much an individual has 'the right to know.' This is encapsulated quite well by Kiritani, an American living in Japan, in her newspaper column *My Nippon* (Kiritani 1997). She describes the difference between airline announcements to a packed 747 plane before takeoff. The English announcement explicitly stated "The test shows that there is a problem with the engine and fuel injection system on the left side of the aircraft. Since we have a long trip ahead and safety is our first concern, the engine will be repaired. Our projected flight time will be two hours from now." The Japanese announcement was terse "Due to unforeseen circumstances there will be a two-hour delay. Please make yourself comfortable. We apologize for the inconvenience." A Japanese businessman sitting next to her understood both announcements and simply chuckled and said, "I'll never understand why Americans always have to know 'why'".

This 'vagueness' is not used in a pejorative sense here. Japanese, of course, is not the only language that uses vague language. As Channell states in her 1994 book (p. 197): "Vague language is not bad or wrong, but nor is it inherently good. Its use needs to be considered with reference to contexts and situations, when it will be appropriate, or inappropriate."

However, the 'vagueness' of Japanese seems to differ somewhat from the "vague language" of Channell (1994: 20):

> For the purposes of this book, I shall work with the following definition of vague language and its use. An expression or word is vague if: it can be contrasted with another word or expression which appears to render the same proposition; it is 'purposely and unabashedly vague'; its meaning arises from the 'intrinsic uncertainty' referred to by Peirce [A proposition is vague where there are possible states of things concerning which it is intrinsically uncertain whether, had they been contemplated by the speaker, he would have regarded them as excluded or allowed by the proposition. (Peirce 1902)].

As shown by the examples given above and discussed below, these premises are difficult to apply to Japanese because there is often no suitable alternative word or expression nor is there any intent (to have a plan or design in mind for the choice of word or expression; i.e., it is often uncontemplated) at vagueness.

Channell's work, however, is illuminating in that the summary at the end of her book offers some hints as to why such a difference in 'vagueness' might occur. She lists twelve items that emerged from her work and items 9, 10 and 11 are of particular interest in the present discussion (Channell 1994: 197-198):

> Vague expressions provide evidence for a 'language of thought'. The expressions may be used to attempt to express thought and concepts for which the speaker lacks appropriate precise vocabulary, either because it does not exist, or the speaker does not know it. For these expressions, it is apparently impossible to describe their meanings independently of consideration of context and inference. Understanding them requires hearers to bring to bear not just knowledge of the lexis and grammar of English, but also pragmatic knowledge about how language is used, and how it relates to its settings.

2.9 Different purposes for language and rhetoric

Channell's pointing to the importance of context might be a key to the 'vagueness' that is often attributed to the Japanese language. The very basis of language usage seems to differ between Oriental cultures (at least, Chinese, Korean and Japanese) and the West. This is pointed out by Connor (1996: 21-22) referring to the work of Carson (1992): "Japanese children learn to understand the value of language in this context as expressing social cohesion, not as primarily a medium for individual expression". Carson's research also included examination of Chinese culture, with a different conclusion: "As in Japan, language is not thought of as primarily a medium for expressing individual meaning, although in China clear public expression is valued as a tool for successful communication" (quoted in Connor 1996: 22). Connor (1996: 24) also goes on to comment that genre-specific studies of contrastive rhetoric, an improvement over earlier studies which did not specify genre, do agree with the traditional findings, including the observation that the Japanese and Chinese tend to be more indirect than Americans.

From the field of professional discourse, Dennett (1988) writing an article on "How rhetorical structure reflects cultural context in Japanese-English technical writing" for the IEEE Transactions on Professional Communication, quotes Ruch (1984: 235) as saying that the "purpose of communication in Japan is to transmit information and provide an emotional message. Nonverbal communication is an important part of this message." Dennett (1988: 119) concludes by warning that "as in the case of Japanese ESL writers brought up to admire the quasi-ambiguity and emotion-evoking model of *kishotenketsu*, it may be necessary to address the cultural expectations that underlie rhetorical models as well".

A Chinese graduate student in communications wrote about his experience in TESOL Matters, a newsletter of the Teachers of English to Speakers of Other Languages. Chen (1997) contrasted the "very direct and formulaic" U.S. English writing with that of Chinese writers who "prefer indirection and rely on metaphors to present their ideas" (Chen 1997: 13):

> For Westerners, rational individuals explore and study the universe, using scientific methodologies to discover the concrete; they search for truths by seeking rules and reasons, structures and details. For the Chinese, in contrast, the universe represents a truth that human beings are already a part of; humans do not possess the nature to discover the secrets of every piece of the world, but they can achieve a harmonious state in which human beings become one with the mysterious, universal way [...]. This difference in perspective suggests why the Chinese and the English languages and writing styles differ so much. (Chen 1997: 13)

This purpose of language and rhetoric differing between China and the West was also pointed out by Bloch and Chi (1995). While noting that Chinese rhetoric is not "monolithic" but as complex and continually changing as that in the West, they do note the strong influence of the Confucian classics which were memorized to pass the civil service examinations until the beginning of the 20th century. The aim of Confucian rhetoric was to create "harmony within a society at every level of social interaction", which contrasted with that of Greek rhetoric which was to persuade the audience. They call for more study of contrastive rhetoric in order to help students trying to learn English as a second language.

This importance of social harmony is clearly stated in the writing of the well-known Japanese sociologist Chie Nakane who describes how important terms of address are and how a single word can express so much (Nakane 1970: 27):

> Let us imagine, for example, the case of X, once a student of Y, who, fifteen years afterwards, becomes a professor in the same department as Y and thus acquires equal status. X still addresses Y as *sensei* and will not refer to him as *doryo* (colleague) to a third person. Y may address X as *kun*, treating him, that is as *kohai*, even in front of X's students or outsiders: Y has to be most broad-minded and sociable to address X as *sensei* in such a context (i.e. the English usage of Dr or Professor). It may also be that when X is well known, but Y is not, Y may intentionally address X as *kun* in public in order to indicate that, 'he is superior to X, X is only his *kohai*'.

This importance placed on social relationships seems to override what might seem to be 'rational' language usage. The lack of rationality pointed out by Nakamura sits at the very heart of the language which is evident from the lexis itself (Nakamura 1964: 532):

> The original Japanese language, as clearly revealed in its classical literature, has a rich vocabulary of words denoting aesthetic or emotional states of mind. On the other hand, words denoting intellectual, inferential processes of active thought are notably lacking. In the original Japanese language, where words were for the most part concrete and intuitive, the construction of abstract nouns was lacking. Hence it is extremely difficult to express abstract concepts solely in words of the original Japanese. When Buddhism and Confucianism were later introduced to Japan and philosophical thinking developed, the vocabulary which was the means of expressing these philosophical thoughts was entirely Chinese, written in the same way, but pronounced differently. Although Buddhism was so widely propagated among the people, its scriptures *were never translated* into the Japanese language.
> "In our country, there is no attempt to translate [Chinese versions of Buddhist scriptures]" said Kokan Shiren (1278-1346), in his *Genko shakusho* (a history of Japanese Buddhism), and he cited this fact as a characteristic of Japanese Buddhism.

2.10 'Thought' without 'thinking'

What is being described is a very interesting example of how 'thought' can be introduced without 'thinking'. Continuing his explanation of the fact that the Buddhist scriptures written in Chinese characters were accepted by the Japanese without their being 'translated' into Japanese, he hypothesizes: "What was the situation, then, after philosophical theories were introduced from the Chinese continent? It seems that Japanese scholars who first acquainted themselves with these theories were so hard pressed merely in learning to use Chinese ideographs that they did not get to the point of understanding and assimilating the thought expressed therein" (Nakamura 1964: 539).

This may be a very interesting example of how a genre itself can assume a meaning of its own. While the Chinese translated the Buddhist scriptures when they were introduced from India, the Japanese merely accepted them because they could 'read' the Chinese *kanji* characters. However, their reading of the ideographs differs from the Chinese and the syntax is different. Despite these facts, the scriptures were acknowledged by the Japanese as being sacred scriptures and allowed to play important roles in their personal and social lives. Would this not be evidence that a genre can undergo partial transformation (different form) when introduced to a culture and yet retain some of its genre features (similar substance and action) to enable it to still belong to a global family of that genre? In the Wittengensteinian sense, there would still be a family resemblance; in the Zadehian sense, it would still be in the same fuzzy set.

2.11 The possibility of alternative forms of a genre

If genre is considered to be composed of substance, form and action, and the form features can be segregated into those claimed by the language community and those claimed by the discourse community, then it should be possible to have a genre in another language with features similar ('same' would not be possible because a different linguisticulture would necessitate a difference) enough to warrant it being categorized together with another genre in a different language. In other words, this would support Mauranen's (1993b) suggestion that genre might be on a higher semiotic plane. Genre can be considered to have moved a step away from the constraints of the register or lexicogrammar of a single language form. In the case of the research article, its birth and nurturing took place in an English-language discourse community (Bazerman 1988, Gross 1990, Atkinson 1992), and therefore it would assume features characteristic of that language during its development to a point where the research article itself assumes a 'meaning' of its own. (The implications of 'overdevelopment' will be discussed further in a later chapter.) This would suggest that there might be research articles in Japanese, Swahili or Pilippino if there were a discourse community to support such a genre. In fact, there are Japanese research articles [though they seem to have become extinct in some fields and have also been influenced by English writing techniques (Kinoshita 1981)] and it seems that efforts are being made to cultivate modern scientific languages in Swahili, Arabic, Bahasa Melayu, Hebrew and Pilippino (Swales 1997).

The question may also arise of whether science, if it is indeed a reflection of a Western approach to science/knowledge, can be written about in a non-Western language. The richness of Japanese scientific discourse may in fact be the reason for the problems the Japanese seem to have with learning English – there is no pressing need for English until the possibility arises of participation in an international discourse community. In other words, all university courses and academic discussions (spoken and written) on all topics can be conducted in Japanese, albeit with the ample use of loanwords for technical terms and concepts. Therefore, unless an international audience is present, most academic conferences in Japan are conducted in Japanese. As evidence for the vigor of Japanese research, Nature Japan K.K., which is the Japan representative of the prestigious British scientific journal *Nature*, announced that the number of scientific papers contributed by Japanese researchers to that journal doubled in 1998, with a significant rise in those related to physics (*Yomiuri Shimbun*, May 20, 1999: 3). This number was "almost on a par with other European countries" although it was far below the papers written by American and British researchers.

2.12 A need for genre evolution in science discourse

The appearance of a genre in a different linguisticulture does not mean that the English rhetorical styles would have to be slavishly reproduced. Rather, as explored in Chapter 1, by segregating the various expectations of a genre, its functions can be demystified and carefully examined in preparation for use or adaptation. Only after knowing what it is you are dealing with can you try to modify it to suit your needs. In traditional Japanese teaching, there is the concept of *shu-ha-ri* which expresses the steps in mastering an art. *Shu* means 'to accept', which suggests that the first step is to completely accept the 'rules' as they are taught by a master. After this has been done, the student is ready for *ha* which means 'to break out' from or 'to modify' what he/she has learned. *Ri* means 'to separate' or 'to create'; after the student knows all the rules and exceptions, then he/she is ready to set out on a path of his/her own. Although there are growing concerns about the overemphasis of form in ESP (English for specific purposes) teaching and calls for the acceptance of alternative rhetorical styles in international academic discourse (Swales 1997, Dudley-Evans 1997a), it would be naive to tell a non-native speaker to attempt instigating a revolution without first having mastered what the discourse community expects. Only after this can there be a breakaway and the creative adoption of an alternative rhetoric.

While the ideal situation might be to have the discourse community itself offer to change and accept alternative rhetorics, under the present circumstances, this is unlikely to occur in the immediate future. Many have expressed caution about relying too heavily on one type of rhetorical culture. Mauranen, in concluding her treatise on *Cultural Differences in Academic Rhetoric* states (1993a: 263):

> Dominant cultures usually dictate the norms of good writing, which is a fact of life for many Finnish academics: in order to get published, it is necessary to adopt the language and rhetorical practices of English-speaking culture. This may not (*sic*) a desirable state of affairs for the academic world. Not only does it disadvantage academics in minority cultures, but it may also lead to reduced variability in academic writing more generally. If the rule of Anglo-American writing norms leads to excessive standardisation of research reporting, it may also have consequences which limit variation and innovation in the research

itself. The coexistence of several rhetorical traditions in the international academic community is probably healthy for research.

Swales (1997) heralds the advent of a geolinguistically changing world. In advanced writing classes for international ESL students at the University of Michigan, for example, in research paper writing, rather than teaching the genre conventions, he and his students examine academic traditions from around the world to explore and experiment with "the soft underbelly of the academic process" which is just as important as "the hard carapace of its polished product" (Swales 1997: 381).

Dudley-Evans (1997a), who has had considerable experience with international students, raises questions about the single-minded use of genre models and rather calls for "the transfer of national rhetorical styles to publication in English" (Dudley-Evans 1997a: 357), but also acknowledges "the complicated system of reviewing and the desire to preserve the standing of the journal may lead to a lack of willingness to accept for publication manuscripts that follow a very different rhetorical style from the norm." However, he urges the demystification of academic discourse, the awareness of intercultural differences, and consideration of how alternative rhetorical styles might be allowed publication in English.

These are serious statements indeed, in particular that of Mauranen (1993a), which goes as far as suggesting that further development of Western science might be hindered by its own success. As an example, let us consider the genre of the science research article as possibly limiting the kind of research being conducted. Its form is efficient for presenting studies done using the experimental scientific method which can be traced back to Boyle and Newton (Atkinson 1992), but does not do justice to what can be discovered through qualitative research. Thus, there may be a need to develop efficient ways to persuade discourse communities about the results of such studies using alternative methods. Edge and Richards (1998) offer suggestions for the disciplines of TESOL and applied linguistics in their article "May I see your warrant, please?: Justifying outcomes in qualitative research." That they felt the need to publish such an article suggests that there is a need for other types of research and that the genres for expressing them have not been as well developed as the (experimental) research article genre.

2.13 Searching for a way to introduce change

Summarizing what we have considered in this chapter, we find that cross-cultural differences in linguisticultures can be a problem. For a non-native speaker of English trying to gain a voice in a focused global discourse community, these 'invisible' differences, in addition to the more obvious ones of surface forms, can be a source of great frustration.

If the differences in perspective are as great as suggested by Chen (1997) and other researchers in cross-cultural rhetoric (Kaplan 1966, 1987, 1988, Connor and Lauer 1988), then as Mauranen laments, the "writer may appear incoherent or illogical" or "may seem exotic and command low credibility" (1993a: 263). Even the editor of *Science* states that "If you see people making multiple mistakes in spelling, syntax and semantics you have to wonder whether when they did their science they weren't also making similar errors of inattention" (Gibbs 1995). This shows that there are people who might equate language usage with science!

As suggested by Swales (1997) and Dudley-Evans (1997a), what would be very interesting is if alternative rhetorics could find ways of gaining entry into professional global genres. As a genre is a complex entity which is both shaped by the discourse community and shapes it at the same time, I do not think alternative rhetorics can be adopted per se, but perhaps some form of contribution could be possible via a genre still under development.

There are genres in other fields where differences are celebrated. For example, in the translation of literature that comes to the attention of those who chose the Nobel Prize winners for literature, the prose may be presented in English but the rhetorical patterns and images are peculiar to the culture and language from which they arise. They are not rejected for this but rather acclaimed for presenting new vistas. Of course, academic discourse is a different type of language but it would be mind-broadening to allow some of the diverse rhetorical styles of various languages into shaping the genres of scientific discourse. By modifying the form, this might even lead to modification in the substance and action. At a time when some are pointing to the "end of science" (Horgan 1996) or question the writing of fellow scientists (Humphries 1998), this might be a way toward new vistas.

Another argument for accepting more diversity in scientific discourse would be to prevent the 'Matthew effect' so-named by Merton in 1968 when he "noticed that in science, credit for a discovery tends to go to the most famous researcher associated with it rather than to the most deserving one". In a classic paper in *Science*, Merton dubbed the phenomenon 'the Matthew effect', after a verse from the biblical book of Matthew: "Unto every one that hath shall be given… but from him that hath not shall be taken away even that which he hath" (Gibbs 1995: 82). If one of the basic reasons for publishing is to gain ownership rights, this concern for fair recognition should not be ignored.

Connor (1996: 26) in her book on contrastive rhetoric calls for us to "celebrate diversity" and quotes McKay (1993a, 1993b) on power relationships in the classroom, which could eventually lead into the academic world. McKay calls for a rethinking of one-way initiation into the academic discourse of Western universities and the "validation of other intellectual traditions that can inform academic discourses".

Perhaps now is the time to think more about ways in which a rich variety of rhetorics or discourses can be utilized to expand our vistas of the world around us, or what we consider to be science. In order to realize change and introduce diverse viewpoints, two factors must be considered: the defense mechanism of the community that resists change (Swales 1998) and the means of implementing change. One of the best ways to penetrate a defense system is to gain entry into the host itself. In this case, the best way to change a community is to become a member of it and then contribute to its development. One obstacle to membership in the case of a discourse community is the language and its usage, which can pose the various issues discussed in this chapter. Therefore, an NNS wishing to gain entry into a discourse community must first learn its rules and expectations *(shu)* to demonstrate the competence to be accepted as a member before being able to break away from these binders *(ha)* and offer differing views *(ri)*. The second factor to consider is how change can be implemented. This book suggests that the genre of the review article offers a place where an opportunity exists to introduce different rhetorical styles and therefore different worldviews. This opportunity arises from the nature of the review article itself and the fact that it seems to be a genre still under development.

3. The review article and its place in the discourse community

> Reviews are comparatively short statements of the status of a scientific problem or idea. These papers are ordinarily solicited [...]
>
> (*Proceedings of the National Academy of Sciences of the United States of America* 1995)

3.1 General considerations

This chapter will reveal the review article to be a very important genre in science that is used to synthesize and evaluate knowledge in a field and also serves as a 'bridge' genre, or a kind of text that functions in more than one discourse community.

The review article, unlike the research article which has been well studied (starting with Swales 1981), is difficult to define. However, this is what this chapter will attempt to do. In order to succeed, an insider's view of the discourse community is essential. This is because text genres are used by discourse communities to synthesize worldviews. Therefore, the entity being defined would not exist without a discourse community. Furthermore, this book is trying to identify a 'live' and evolving genre in a dynamic community in order to find ways to enable non-native speakers to use the genre. This requires not only examination of the written 'rules' of this genre, which can be obtained from instructions and the texts themselves, but also discovering the expectations of active members of the community itself. As discussed in Chapter 1, Collins (1982) pointed out the weakness in early sociology of science studies that examined only data about the community. Also, Berkenkotter and Huckin (1995) have stated that "to date, very little work on genre in rhetorical studies has been informed by actual case research with *insiders*" (1995: 2; emphasis in original). To prevent these shortcomings, let us begin by examining the written 'rules' pertaining to review articles and also consider the opinions of specialist informants, or insiders in the community.

As a first step, Swales' comment is helpful: "a discourse community's nomenclature for genres is an important source of insight" (1990a: 54). In practice, the scientific discourse community uses many names for the review article – review, review article, general article, and perhaps in keeping with the rapidity of developments, minireview, which is a brief summary of "developments in fast-moving areas" (*Journal of Bacteriology*, 1995). This lack of consensus with names and, as we will see below, dearth of instructions on its form, may be an indication that this genre is still under development and thus may offer an opportunity for more genre freedom.

This 'looseness' is in contrast to the research article which is often clearly defined as to format and structure by many journals (e. g., *The Biochemical Journal, Analytica Chimica Acta, Clinical Chemistry, European Journal of Biochemistry*), and has been the target of study by many applied linguists (Gunawardena 1989, Hopkins and Dudley-Evans 1988, Swales 1990a). Perhaps because of its importance as a genre in the science discourse community, there are materials for teaching how to write research articles, as has been discussed in Chapter 2. However, as presented in the next section, there is very little discourse community consensus available in written form about the review article genre.

3.2 Journal instructions

3.2.1 Genre name

Before attempting to define the review article, let us examine what the journals themselves say about articles that are not reporting original research but are aimed at presenting overviews or analyses of the field or part of it or synthesizing available information. The name of the genre itself varies. Random perusal of the instructions for contributors

(1995) reveals that although the name 'review' is most common (e.g. *The Biochemical Journal, Analytical Biochemistry, Analytica Chimica Acta, Chemical Society Reviews, Clinical Chemistry*), names such as 'minireview' *(The Journal of Biological Chemistry, The Journal of Cell Biology, Journal of Bacteriology)* and 'commentary articles' *(The Journal of Immunology)* are also found. *The Journal of Cell Biology* accepts both mini-reviews and commentaries and distinguishes between them as follows:

> Mini-reviews should not contain an exhaustive review of an area, but rather a focused brief treatment of a contemporary development or issue in a single area [...]. Commentaries should offer a more personalized perspective or synthesis of information on a topic that will be of interest to the general readership.

Analytical Chemistry (1995) proclaims that "only a year ago the journal underwent a complete facelift. We developed several new sections." Interestingly, one of the new developments is called an "A-page section" which carries report articles described as follows:

> Report articles are intended to broaden readers' professional interests and keep them aware of the role of analytical chemistry in the scientific arena and in society in general. Suitable subject matter includes, but is not limited to, state-of-the-art coverage of important, broad techniques; historical views of certain facets of the field; tutorial presentations of newly developing technical areas; applications of techniques to areas of topical interest, such as environmental chemistry and biomedicine; and philosophical presentations dealing with any facet of analytical practice.

This description is very close to what is later presented in this book as a 'review article.'

The continuing development of genres is well expressed by the journal *Biochemistry* in its instructions to authors (1995): "we are dealing with a moving target, as what may have been quite acceptable five years ago may no longer be appropriate. For example, the expectations for NMR [gloss: nuclear magnetic resonance] structural studies have changed significantly over the past few years as detailed in the new Instructions to Authors" [overall instructions, not specifically relating to reviews].

3.2.2 Authorship

One significant difference between research articles and reviews is that the latter are, as a rule, commissioned by the journal. While submission of research articles is usually open to all members of the relevant discourse community, the opportunity to write a review article is much more restricted but once the invitation is extended, the writer is usually allowed more freedom, such as with respect to length and structure. *Nature*, in its instructions to contributors states that, among other articles, such as progress articles, articles (research reports), letters to *Nature* ("short reports of outstanding novel findings"), it publishes review articles. Interestingly, while the length limits are given for the other categories, the instructions for review articles states that the "length is negotiable in advance". It also states that although reviews are commissioned, suggestions for possible contributions can be made by sending a one-page summary to the Reviews Coordinator. This requirement of prior consultation with an editor before submitting an unsolicited review also holds for other journals, such as *The Biochemical Journal, Analytica Chimica Acta, Clinical Chemistry, European Journal of Biochemistry.*

Science, published weekly by the American Association for the Advancement of Science, has 'general articles' of 3,000 to 5,000 words (three to five printed pages), which "review new developments in one field that will be of interest to readers in other fields; describe a current research problem or technique of interdisciplinary significance; or discuss some aspect of the history, logic, policy, or administration or science. Readers should be able to learn from a general article what has been firmly established and what are unresolved questions or future directions" (*Science* 7 Jan 1994, Vol. 263, pp. 37-40).

Before examining this more closely, let us see how some international journals view the review article. This can be done by examining the Instructions to Authors of various reputable international science journals.

> *The Biochemical Journal*, Vol 305, part 1, 1 Jan. 1995
> Reviews will usually be solicited, although unsolicited reviews will be considered for publication. Prospective writers of reviews should first consult the Reviews Editor, via the editorial office, and should enclose a short (one typed page) summary of the area they proposed to cover.

Analytica Chimica Acta (An international journal devoted to all branches of analytical chemistry), Vol. 297, No. 1-2, Oct. 31, 1994
Reviews are normally written by invitation of the editors, who welcome suggestions for subjects.

European Journal of Biochemistry, Vol. 233, No. 1, October (I) 1995
Authors wishing to submit a review should contact the Reviews Editor [...]
[*Note: Nothing else about the review.*]

3.2.3 Summary of journal expectations

Examination of the instructions to authors from the various natural science journals given above reveals the following:

- The science discourse communities covered recognize the existence of a genre of 'review' which is sometimes called 'mini-review' or 'report article'.
- Although rather detailed information is given about the features of a 'research paper', there is not much said about the 'review'.
- 'Reviews' are usually solicited by the editors and unsolicited ones may be acceptable but seem to require editorial acceptance prior to formal submission.

There are some journals that are especially dedicated to reviews. Here is one example.

Chemical Society Reviews, Vol. 24, Issue 5, inside cover
It is intended that Chemical Society Reviews will have the broad appeal necessary for researchers to benefit from an awareness of advances in areas outside their own specialities. Deliberate efforts will be made to solicit authors and articles from Europe which present a truly international outlook on the major advances in a wide range of chemical areas. It is hoped that it will be particularly stimulating and instructive for students planning a career in research. The articles will be succinct and authoritative overviews of timely topics in modern chemistry. [...] In general, authors, who will be recognized experts in their field, will be asked to place any of their own work in the wider context. [...] Although the majority of articles are intended to be specially commissioned, the Society always considers offers of articles for publication. In such cases a short synopsis (including a selection of the literature references that will be cited in the review and a brief academic CV of the author), rather than the completed article, should be submitted to the Senior Editor (Reviews) [...]

In this case, not only is an outline of the proposed review required but the curriculum vitae of the author, which indicates the emphasis placed on the level of experience required of someone writing a review article.

To summarize, the science discourse community views a 'review' as a usually solicited article, which is longer than a research article and covers the state of the art, presents a historical perspective or discusses a current issue in the field. It is usually commissioned by the editor of the journal and thus, written by a researcher established in the field.

3.3 Specialist informants

3.3.1 Rationale for consulting specialist informants

In this work, the discourse community is considered to be the basis of a science worldview. Thus, the views of the people composing the discourse community itself are important when trying to understand genres. As discussed in defining the discourse community and genre, we need to look inside the discourse community in order to learn about genres – their structure, form, and action. The importance of this is stressed by Berkenkotter and Huckin (1995) who note that Bakhtin considered it essential that genres be studied in their actual social contexts of use. In this book, we have defined science as being a dynamic worldview held by a focused global discourse community, which maintains that view as a metastable equilibrium reached by intercommunication through language used as genres. Therefore, to understand a genre, we must turn to the 'insiders' of the discourse community to understand it.

The idea of consulting field specialists was strongly advocated by Bley-Vroman and Selinker in 1984 in their five steps for "an optimum research strategy". However, Swales (1990a) discusses the problems which can arise from too much reliance on these specialists; the specialists themselves have their own biases, the work is time-consuming, and the findings can be equivocal when different informants are consulted. Continuing this discussion on specialist informants, Hozayen (1994) comes to a conclusion, echoing Swales (1990: 130), that they are needed in "a partial and supportive role". She used Ph. D. medical students who were working as assistant lecturers and a university English language teacher who had been a general medical practitioner to validate her analyses of the discussion section of medical research articles. She had them guide her to key texts which others in the field would be reading and also had them validate the moves which she established for the discussion section.

Indirect support for using informants from the text analyst's viewpoint comes from the work of Hunston (1993) on evaluation and ideology in scientific writing in which she points to the difficulties of learning how to write for a discipline. Hunston states (1993: 72): "The final product must be expressed in a way that both says what the student wants to say and fits what the reader expects to hear. The ideology of the description must be conformed to, yet its value system must remain implicit". This implicitness can be invisible to an outsider and has been expressed by Swales (1990a: 130) as follows: "the general need for a specialist informant may be in inverse proportion to the degree of relevant experience possessed by the genre analyst".

This knowledge of the genre analyst or language specialist has been discussed by Dudley-Evans, who has worked with specialist informants for team-teaching on lectures and examination answers (Johns and Dudley-Evans 1980) and for the instruction of more advanced research students in relation to research project outlines and the writing of titles, dissertations and essays (Dudley-Evans 1986, 1995). He states that various factors affect such collaboration work and that mutual respect is important. He also points out that the language specialist should have an understanding of the basic concepts underlying the materials under consideration for analysis or teaching.

For the present work, it was decided that specialist informants would be valuable for gaining information on the insider's view of the review article and the corpus of articles, and this was found to be true as will be discussed below. Another reason for consulting specialist informants was similar to Hozayen's (1994) purpose, to validate the text analyses. Unfortunately, this was not very successful as will be discussed below.

3.3.2 *Specialist informants consulted*

As specialist informants, members of science, engineering and medicine departments at the University of Birmingham were asked for co-operation in examining review articles in their respective fields of specialty (see Appendix 2). These scientists were recruited by approaching someone in the relevant department and asking for an introduction to those in the department who were conducting research in the field of the review articles under study. The priority was field relevancy over seniority but all informants except one were staff members of lecturer status or above. The initial contacts were made by direct meeting or via e-mail followed by meetings once to explain the purpose of the study and what was requested of the informants and sometimes a second time to personally receive comments after they had had time to consider the article assigned. In some cases, the responses were returned by campus mail.

3.3.3 *Questionnaire responses from the specialist informants*

Eleven specialist informants responded to questionnaires about review articles, the *Proceedings of the National Academy of Sciences of the United States of America*, and the review article(s) they were examining. All agreed that reviews are important – from 'very' to 'fairly.' The difference may have arisen from the differences in the fields. Those who replied 'very' were in the rapidly developing fields of biotechnology, biochemistry, liver growth, viral pathogenicity, plant physiology, and plant virology. The informant who replied that reviews were only 'fairly' important was in botany, specializing in taxonomy. Two mentioned the usefulness of reviews as teaching materials.

The specialist informant responses presented the following profile for review articles: they are generally considered to be of moderate to high importance, are useful for their usually comprehensive literature references, and are helpful for teaching students. Reviews are read to bring the informant up to date in their field, including information from papers with which they may not be familiar, or to learn about an unfamiliar field with which their interests may be associated. They are also useful as convenient teaching materials which someone else has prepared.

When asked who writes review articles, the eleven informants all agreed that the writers are 'experts' who are 'distinguished' or 'established' in their field. This matches their comments during the interviews. Two did say that sometimes young researchers are also asked to write one. Of the eleven, nine had written at least one review.

When asked why review articles are written, most gave the expected answer of summarizing information or 'reviewing' the field. Other responses were to point out "new avenues of research" and also to present controversial or overlooked areas and also to attempt to resolve controversy. One interesting response was "to sell review journals". This informant said that he often used review articles to teach undergraduates and, as noted in the interview section, said that he was aware of at least two instances in which a textbook was published shortly after the review.

Another interesting set of responses were related to the author's reputation. Four of the seven informants pointed out that writing a review was an opportunity to enhance one's reputation and publicize one's work. These responses indicate a prestige factor of being invited to write a review – the editors recognize the importance of the writer in the field of specialty. Here is an example of the wording from a letter of invitation suggesting that being asked to write a review can be considered to be an accolade: "It [title of work was given in the previous sentence] will have a compilation of review articles pertaining to various aspects of lipids research authored by prominent scientists from all parts of the world" (letter from S.G. Pandalai, Managing Editor, *Transworld Research Network*, February 7, 2000).

3.3.4 Interviews with specialist informants

In addition to the above findings, the interviews with the specialist informants revealed the following: (1) the pace of progress in science field is extremely rapid; (2) there are various types of reviews; (3) well-known researchers in the field are invited to write reviews; and (4) reviews are often cited, suggesting that these texts have an important impact on the construction of knowledge in their respective fields.

3.3.4.1 'Explosion of information'

One significant feature of reviews mentioned by many of the informants, although not necessarily specific to reviews, was the 'explosion of information'. Many pointed out that the field being discussed was rather 'new', having begun only three to ten years ago. Professor Robert H. Michell pointed out that the last paragraph of the review he had examined (which was only about two years 'old') was already out of date. Senior Lecturer John Newbury explained the numerous instances of reference to 'personal communication' in one of the articles as being due to the author wanting to present the latest developments in a laboratory from which he had recently moved. The author did not seem to want to wait the 6 to 9 months it would take for the information to appear in published form. Dr. Phil Lecane, who had recently completed his Ph.D. thesis, complained about the difficulty of writing it up because of the literal flood of papers in his very specific field over the past three years.

3.3.4.2 Different types of reviews

These informants also pointed out that there are different kinds of reviews. First, they can be divided into those for a 'general' audience and those for experts. The former would be useful for teaching and helping young people "get some feeling for how a field came about" (Michell). "A very clearly written undergraduate essay" is how Senior Lecturer Ian Mitchell described the SVAU article. As mentioned earlier, Lecturer Richard N. Lester pointed out that he knew of at least two instances in which a textbook followed the publication of a review article. When presented with the SCLE paper, he immediately searched for reference to a seminal paper in this area, which had changed the entire field of plant classification. The review article was also considered by Senior Lecturer Clive Sweet to be a good teaching instrument. In addition, for a researcher entering a new field, the 'general' review article would be a good starting point for learning the basics and noting the important field literature.

The other type of review article would be a more state-of-the-art type which a researcher would probably read rather regularly to keep up to date on recent developments and literature. Senior Lecturer

Alastair Strain even points out their usefulness for finding names and addresses to gaining access to reagents. Senior Lecturer Kenneth Roger Wood noted a basic difference between the two reviews he was asked to analyze – one being useful for both researchers and teachers and the other covering only a very limited area in which the author was trying to propose a theory.

These observations suggest that perhaps there are also several types of 'expert' reviews. Here Michell's comments were illuminating. He specifically stated that different types of reviews are written at different times. In his field of cell signalling, there was an explosion of information in the 1970's which was followed by a need for someone to summarise and synthesise it into models which could be used to predict, to lead to new experiments, and to move the field forward. Professor Chris M. Thomas's comment was also interesting in that he stated that the HOLS article was not a simple historical re-evaluation of simple reiteration but one that actually went back to the original literature to "put the record straight." Related to this comment was the explanation by Mitchell, working in psychology, that one of the references in the paper he was asked to examine had led to the author being treated as an outcast when it first appeared – the fact that it was included in the review indicated that it was now accepted as an important reference in the field.

3.3.4.3 Experts as authors

The review article is said to be written by well-known researchers in the field. For example, the author of HKNU had written an earlier seminal paper, the author of SCLE was a member of a department which had been voted by peers as the best one in the field, the author of SWIL would be one of the names to "look for" when searching databases, and the author of ITEM was one of the discoverers and world's leading experts on the topic. The specialist informants also repeatedly mentioned looking for certain names in the references before actually reading the articles.

3.3.4.4 Cited half-life

Lecturer Rhulon Alfred Thornhill pointed out that the *Proceedings of the National Academy of Sciences* is considered to be an excellent journal from the aspects of both impact factor and cited half-life. According to the *Journal Citation Reports (JCR)*, the impact factor is "a measure of the frequency with which the 'average article' in a journal has been cited in a particular year" and the cited half-life is "the number of journal publication years going back from the current year which account for 50% of the total citations received by the cited journal in the current year" (ISI 1993: 10). The impact factor can also be calculated for a journal "by dividing the number of all current citations of source items published in Journal X during the previous two years by the number of articles Journal X published in those two years" (ISI 1993: 10). A high impact factor and a long cited half-life would suggest that an article is an important one. *JCR* also states that "Reviews are often more highly cited than original research articles" ISI (1993: 16). As being cited more means having a greater influence on the construction of knowledge in a field, the review seems to be an important genre.

3.4 Definition and propositions for the review article

3.4.1 Definition of the review article

The information from the various journal instructions related to reviews and the interview and questionnaire responses from the specialist informants led to the conclusion that in the science discourse communities examined in this study, review articles:

(1) Are written by established authorities;
(2) Are usually commissioned by journal editors;
(3) Vary greatly in length and style;
(4) Address different audiences – generalist (researchers entering a new field, teachers, students) or specialist (peer researchers);

(5) Come in different types which may be related to the extent of progress in the field – describing the state of the art, presenting historical perspective, pointing out current issues, synthesizing theories or models;
(6) Are well-read and well-cited.

The fact that the review article has many names and comes in different types may raise the question of whether or not these texts constitute a single genre. However, the findings presented above suggest that these texts do constitute a recognizable genre for two reasons. First, all of the specialist informants recognized the review article as a specific type of text although they did note that it could come in various types, perhaps suggesting the existence of subgenres. The second reason is that the *Journal Citation Reports* (Institute for Scientific Information, 1993: 7) recognizes the following types of articles: original research articles, review articles, and technical notes or short articles. It states (ISI 1993: 16): "articles are coded as reviews if they have more than 100 references, if they appear in review publications or review sections, if they have the word 'Review' or 'Overview' in their titles, or if the Summary states that the article is a review or a survey".

According to the definition of 'genre' in this book (section 1.10), if members of the community recognize certain texts as belonging to a genre, then this would be sufficient to classify them as such. As shown above, review articles are recognized by both the members of the discourse community (specialist informants) and *JCR* (ISI 1993), which is a unique service aimed at serving the needs of specialist communities, for example, helping them choose where to publish and select which journals to scan regularly to keep aware of current research.

In sum, the review article is considered by its discourse community to be a prestigious, authoritative genre which, perhaps because it is written by established experts in the field, allows more freedom in terms of substance and form than the more rigidly defined research article. Despite these 'fuzzy' elements, the review article can be classified as a genre because it is recognized by the focused global discourse community as medium of intercommunication with specific community-acknowledged purpose(s).

3.4.2 An opportune genre for a new age

Based on the above definition of the review article, I would like to propose that it is an opportune genre with which to introduce new linguisticultures for a new age. The review article is influential in helping shape knowledge within its community, is important for helping community members cope with rapidly advancing research, and is also influential outside of its specialist community.

3.4.2.1 Influence within the discourse community

A very good example of the influence a review article can have is expressed by Bizzell (1992), one of the initial proponents of the "discourse community" concept (Swales 1998: 197). Working in the field of composition studies, she describes how she developed her basic ideas. She states the following (Bizzell 1992: 17-18):

> The concept of "discourse community" thus was initially used to define an alternate research site for scholars in composition studies. Indeed, in my review-essay I presented that research was already going on at that site; as far as I knew, it really wasn't, but I hoped to invoke it with my essay. I don't remember whether the concept of discourse community was "my own idea." I don't really believe that people have their "own" ideas or "own" them. *The formulation of the concept in my review-essay was not taken from anywhere else in its present form, but it was certainly profoundly influenced by my reading in Kuhn, Fish, Richard Rorty, and the sociolinguists, and by discussions with Bruce Herzberg*. Indeed, given the concept's provenance in *a review-essay – which Dominic LaCapra calls the characteristic genre of our intellectual era –* I think it would be fair to say that Flower and Hayes also influence its development. And in turn, *it would seem that this essay has influenced the work of the cognitivists who now define their research as "socio-cognitive," seeming to attempt to build in the supplement I called for*; and *it has influenced the work on discourse conventions* done, for example, by scholars in writing across the curriculum. [emphasis mine]

What Bizzell has done in this one passage is present the dynamic interaction between the individual and the discourse community, recognize the "review-essay" as a genre, and claim that her "review-essay" synthesized various elements to push forward not only the discipline in which she is active but also the work of at least one other. These are precisely the features that we have just seen in the science 'review article'.

3.4.2.2 Coping with information explosion

The review article is likely to become increasingly more important as research output burgeons in the sciences. One of the specialist informants, Lecane in cancer research, described how difficult it was for him to complete the writing up of his thesis because of the explosion of information in his research field on tumor-suppressor genes. This explosion metaphor was also used by Lester for biochemical taxonomy in describing work in chloroplast gene sequencing and Mitchell, in psychology, when discussing a review article on physiological cell death. Michell, working in biochemistry, stated that one of the aims of reviews that he had written was "to offer new, experimentally feasible, syntheses arising from past observations (whenever possible, particularly when a field is becoming detailed but confused)". Note the revealing final comment in parentheses.

This 'problem' of information explosion is expressed well by Paul and Charney (1995: 397):

> Although scientists must read widely to build up shared knowledge, they naturally have practical limitations on their time and attention. Keeping up with science can compete with doing science. Faced with a rapid proliferation of journals published at ever-shorter intervals, scientists adopt rigorous selection strategies to cull a small number of articles that they will actually read, choosing most often those that apply most directly to their current work, and reading only a small proportion of articles completely.

The issue of this blinder-type specialization, i. e., of focusing only on a narrow field of research, was also evident from a comment made by Newbury that it would be difficult to describe "generally accepted phenomena" because this would suggest that "most scientists are aware of them, which may not be the case because of specialisations".

Under such circumstances, there will probably be a greater need for review articles which can reflect upon the research occurring in a field, evaluate it, synthesize different aspects of it, and suggest avenues for future work. For example, Strain, in describing a review article he had written stated that two of the aims were to survey the literature and to discuss future prospects. Sweet stated that the aim of one of the review articles he had authored was "to introduce medical students and other biomedically interested undergraduate [sic] to the poorly studied area of pathogenicity". This accessibility to students was also mentioned by Mitchell, who remarked that the review he was asked to evaluate was "rather like a very clearly written undergraduate essay". This comment suggests that such a review article would be accessible to students, novice researchers and scientists in related fields who may not have the same level of expertise as the specialist.

Another reason for the importance of the review article is that historical re-evaluation is important to, as Thomas said, "put the record straight". He noted that one of the references quoted in the review he was evaluating had not been recognized until another paper appeared which enabled the technique described in the earlier paper to become widely usable. Along a similar vein is Sweet's comment presented above about drawing student attention to the "poorly studied area of pathogenicity". From a slightly different viewpoint, Mitchell stated that one of the aims of a review article that he wrote was "to challenge existing assumptions".

3.4.2.3 Bridging discourse communities

Another interesting notion which arose from the specialist informant interviews and questionnaire responses was that of the review article trying to do more than influence the social action within its native community. The review article was said "to inform a general audience" (Mitchell), "to provide a comprehensive, up-to-date review of all recent work in a particular field (usually for workers in that field but also for workers (or teachers) in related fields" (Wood), and to seem to lead to textbooks (Lester). Even more convincing evidence for the review article serving as a bridge genre is what Thomas, a professor of biochemistry, wrote in response to his aims for writing a review article:

> I have been invited to write a review article on 'The Physics and Biology of Nanomotors' for Journal of Physics D (Applied Physics). This will show physicists how physics can be applied to an important and interesting area of biology.

These remarks can be interpreted as suggesting that the review article can sometimes be purposely trying to address an audience outside of its own discourse community. As quoted above, Bizzell, who works in composition studies, wrote that her review-essay had an effect on the work of cognitivists and on those studying discourse conventions; in other words, it had had an effect on scholars in different discourse communities. These observations indicate that the review article, in addition to its purposes (e. g. evaluating, synthesizing functions) within its own discourse community, also seems to be a genre with which one discourse community tries to affect the actions of another discourse community.

To express this idea of one genre used with purpose (not unintentionally as can occur with historical texts) to affect actions outside of its native community, I would like to use the term 'bridge' genre. This idea may help explain why the review article seems to be very fluid as a genre, for example, in comparison to the research article. Returning to the metaphor of the 'frozen core' representing constancy and 'evolution at the edge of chaos' (presented in Chapter 1) what we seem to be observing here is an evolutionary process still in progress (evidence: the many different names for the genre and lack of 'hard and fast' rules). This might be because the genre itself is trying to expand the very borders of the discourse community and at the same time make an appeal for its global view to other discourse communities.

3.5 Choice of corpus

The review article was chosen for analysis because it is a challenging and important bridge genre, it does not appear to have easily definable features, and it has not been the object of much study. As stated in the Introduction to this book, one of the aims of this study is to find ways to help the nonnative English speaker (NNS) scientist gain a voice in the international academic discourse community in order to promote contributions from alternative linguisticultures. Thus, being able to offer support to write for this important genre is needed, as suggested by the quote above by Sweet that "most academics will attempt to write a review article." In order to accomplish this, the features of the review article must first be more clearly defined. The key to these features, as discussed in Chapters 1 and 2, must come from the review articles themselves. Therefore, the first step was to build a corpus of review articles to be used for the analyses. As such a study of review articles was, to the best of my knowledge, something which had not yet been attempted, I decided to focus on one reputable journal to explore this yet uncharted genre. Let us now turn to the reasons for the choice of the corpus and a description of its features on a macro level before proceeding to the micro level analyses.

In order to obtain reliable information about the review article, care was needed to select articles from a respected journal and yet to refrain from choosing reviews which might be too specialized as this could limit the applicability of the findings. This led to the choice of reviews from a highly evaluated journal with a broad audience – the *Proceedings of the National Academy of Sciences of the United States of America* (hereafter referred to as *Proceedings*). Support for the validity of this choice came from specialist informant information and bibliometric evaluation, which will be explained in the following sections.

As other possible sources, there were prestigious journals such as *Nature* and *Science*, which publish review articles and also more specialized journals devoted to reviews, such as *Chemical Society Reviews*. However, the *Proceedings* were chosen because, in addition to covering a wide range of fields, the 1993 *Journal Citation Reports* ranked it as the most cited journal. Also, unlike *Nature* and *Science*, which are publicly marketed, the *Proceedings* are not, suggesting a more specific orientation of its audience. This audience factor was also the reason that the more specialized review journals were not considered. The audience was expected to be too limited and the subject matter too specialized. As the review article has not been the target of much study yet, it was thought that examining a set of reviews that were from a widely-read journal with a broad audience in focused discourse communities would be appropriate. It is hoped that the findings of the present study can serve as a starting point for further in-depth analyses of reviews.

3.5.1 Bibliometric evaluation

The other main source of support for the choice of the *Proceedings of the National Academy of Sciences* came from its ranking in the *Journal Citation Reports (JCR)*, a bibliometric analysis of the some 6,000 science journals which are in the ISI (Institute for Scientific Information) Database (ISI 1993). It shows how many times a paper (article, note, review, etc.) has been cited in other papers. *JCR* defines a citation as the "formal acknowledgment of 'intellectual debt' to previously published research, publicly recorded in the references listed by contemporary authors" (ISI 1993: 5). *JCR* offers information on the "largest", "most frequently used", "hottest", and "highest impact" journals (as described by *JCR*). In the 1993 *JCR*, the *Proceedings* ranked as the fourth largest journal with 2,441 source items and first as the most cited journal with 234,319 citations.

3.5.2 General description of review articles in the corpus

In 1993, 25 reviews were published in the *Proceedings*. These reviews formed the corpus used for the study and analysis of review articles at the levels of the whole text, section, paragraph, sentence, clause, and lexicogrammar. These aggregations were chosen because they were employed by the writers, whose choices are expected to have been governed by linguistic and discourse community rules. Support for this approach comes from Leydesdorff's work on epistemic networks, who talks about (1991: 75):

> [...] sedimentation of knowledge [...] at various levels of aggregation: at the very base there are words and combinations of words, which add up to sentences. Sentences are organized in paragraphs and sections, which sequence the line of argument within an article. Articles contain the unit contributions ('knowledge claims') of scientific development and form ensembles which constitute volumes of journals, archives of specialties and eventually disciplines.

Appendix 1 gives the complete bibliographic information of the 25 reviews examined. Tables 4, 5, 6, and 7 give the abbreviated name by which they will be referred to hereinafter, the last name of the first author, and the title. The reviews are arranged not chronologically, but in the four categories of history (H), status quo (S), theory (T) and issue (I) type reviews. These categories and the rationale for this classification will be explained below.

3.6 Preparation of corpus for analyses

The 25 reviews were photocopied and scanned into a computer to form a corpus of texts for detailed investigation. Each text was stored as a text file, as an Excel® file with information about its Discourse and Move step classification and as a text file based on the Excel® file. All text examples quoted from these files are presented in the following format:

TAND 1:5	Rx	***This process is called*** potocytosis (9, 10).	Def

The first four capital letters identify the review article with the last three alphabets representing the first three letters of the first author's last name. The first letter identifies the type of review: H for historical, S for status quo, T for theoretical and I for issue. The following two numerals give the paragraph and sentence number, respectively. The R means that there is a reference citation in the sentence/clause (U, unpublished observation; P, personal communication; N, reference with name of author(s) given in text) and x means that this is the last sentence in the paragraph. Multiple x marks (xx, xxx, xxxx) mean that the sentence is the last one in a section, with a greater number of marks used to show the end of a more significant section break, e.g. before a heading or other physical section marker. The bold characters in the sentence/clause represent words or phrases which can be useful as identifying features of the Discourse or Move Step type. At the extreme right is the Discourse or Move Step category in which this sentence/clause has been classified. For complete bibliographic information, see Appendix 1.

HBRO	Brownstein	*A brief history of opiates, opioid peptides, and opioid receptors*
HKNU	Knudson	*Antioncogenes and human cancer*
HNIC	Nicolaou	*Chemistry and biology of natural and designed enediynes*
HOLS	Olson	*The human genome project*

Table 4. Titles of history reviews in corpus studied.

SCLE	Clegg	*Chloroplast gene sequences and the study of plant evolution*
SGOL	Golbeck	*Shared thematic elements in photochemical reaction centers*
SLAI	Lai	*Hepatocyte nuclear factor 3/fork head or "winged helix" proteins: A family of transcription factors of diverse biologic function*
SLEN	Lengyel	*Tumor-suppressor genes: News about the interferon connection*
SRAM	Ramachandran	*Behavioral and magnetoencephalo- graphic correlates of plasticity in the adult human brain*
SSHU	Shulman	*Nuclear magnetic resonance imaging and spectroscopy of human brain function*
SVAL	Vallee	*Molecular analysis of the microtubule motor dynein*
SVAU	Vaux	*Toward an understanding of the molecular mechanisms of physiological cell death*
SWIL	Wilson	*Strategies to protect crop plants against viruses: Pathogen-derived resistance blossoms*

Table 5. Titles of status quo reviews in corpus studied.

TAND	Anderson	*Caveolae: Where incoming and outgoing messengers meet*
TBEN	Bennett	*The molecular machinery for secretion is conserved from yeast to neurons*
TBRE	Breslow	*Transgenic mouse models of lipoprotein metabolism and atherosclerosis*
TGRA	Grant	*Modes and origins of mechanical and ethological isolation in angiosperms*
TPRI	Price	*Alzheimer disease and the prion disorders: Amyloid β-protein and prion protein amyloidoses*
TSCH	Schimmel	*An operational RNA code for amino acids and possible relationship to genetic code*

Table 6. Titles of theory reviews in corpus studied.

IAME	Ames	*Oxidants, antioxidants, and the degenerative diseases of aging*
IBEU	Beutler	*Gaucher disease as a paradigm of current issues regarding single gene mutations of humans*
IBLE	Blenis	*Signal transduction via the MAP kinases: Proceed at your own RSK*
IEPS	Epstein	*The anomaly of silicon in plant biology*
IKEY	Keyfitz	*Are there ecological limits to population?*
ITEM	Temin	*Retrovirus variation and reverse transcription: Abnormal strand transfers result in retrovirus genetic variation*

Table 7. Titles of issue reviews in corpus studied.

The review texts were also subjected to analysis using the concordance program *Wordsmith* from Oxford University Press (authored by Mike Scott, version 2.0). The analyses carried out in this study are listed in Table 8.

3.7 Rhetorical organization and physical features of corpus reviews

The rhetorical organization and physical features were the first levels of characteristics to be identified. As the review texts examined here appeared in a journal, they were assumed to have identifiable rhetorical organization and physical features due to the constraints of the printed journal. These items are summarized in Table 9. The physical features for the beginning included the title and author(s) affiliation. All reviews except one had an abstract, the exception being HBRO which had a quote before the main text. The main body was composed of sections with separate headings for all except one, again HBRO. The end was specifically signalled with a concluding heading in eight articles. The appended items are sections which appeared after the main text itself.

Table 9 summarizes information on the physical components of the 25 review articles. The number of authors range from one to four with 18 of them (72%) having been written by a single author. All except one have an abstract and all except four have an acknowledgements section referring to helpful discussion, comments on the manuscript or financial aid.

Action	Reason
Number all parts of article	Allow quick reference
Identify rhetorical organization and physical features	Obtain quantitative data on physical text elements
Interview specialist informants	Discover discourse community views
Examine linguistic evidence in review articles	Determine what each review is trying to accomplish
Identify moves	Discover information ordering
Identify move steps	Reveal parts of move
Identify SIE	Observe sentence-initial position
Identify GS function	Observe the subject function of the clause
Identify verb	Examine meaning expressed by verb tense
Identify useful words, expressions	Offer help for NNS
Identify key words	Clarify move step features; offer help for NNS
Obtain collocations of useful words, expressions	Offer help for NNS

Table 8. Analyses of review articles conducted in the present study. (SIE, sentence initial element; GS, grammatical subject; NNS, nonnative English speaker)

Rhetorical organization	Physical features
Beginning	Title Affiliation Abstract Quote
Middle	Main body sections
End	Summary Conclusion
Appended items	Note in Proof Acknowledgements References

Table 9. Components of review articles studied.

The overall average number of paragraphs is 30 with the range of 11 for two reviews and a lengthy 65 for one. There seems to be no significant difference in length for the different classes of review articles. In general, there are few figures, the average being 2, and very few tables, with the average being less than 1. Many references are given, ranging from 23 to 217 for an average of 74; the Issue category gives a slightly higher average of 99 references. One of the specialist informants mentioned that a review is a good place to publicize one's own work but the average of references including at least one of the authors averaged only 14.11 %, with a low of 10.13 % for the Status Quo category. Two gave no reference to their own work, however, one article in the Theory category included 36 references to the author's own work out of 66 references, or 54.55 %. This Theory category also averaged the highest percentage of references to the authors' works, 25.21 %, which could be expected because the writers were trying to propose a theory or model of their own. One writer made extensive use of unpublished information, 19 instances (ITEM), while another mentioned 24 instances of personal communication (SWIL). These are not included in the number of references. The average number of printed pages was 6 and the average number of words of text (including the abstract but not the title and headings) was 4,593 for a total corpus of 114,819 words. The shortest article had 1,932 words (TPRI) and the longest had 8,390 (SRAM).

With respect to the headings used in the articles, only one (HBRO) had no headings to divide sections within the article. There was great variation in the types of headings used; one even used all capital letters and centering of each section heading (SWIL). On average, five main headings were used, with the definition of 'main heading' being a phrase set apart from the text itself and used to signal a new section. If the phrase was included at the start of the orthographic paragraph, usually in bold or italic type, and there were other higher level headings, then this phrase was not considered a 'main heading.' When there were no other higher level headings, these headings were counted as 'main headings' because they were the only available signals of section breaks. A specifically signalled concluding section appears in eight of the articles (HNIC, HOLS, SCLE, SRAM, SWIL, TGRA, TPRI, IEPS), and three finish with a view toward the future (TAND, IBEU, SSHU).

Tables 10, 11, 12 and 13 summarize various components of the different categories of review articles studied. Overall, the average number of authors for the entire corpus was 1.64, all except HBRO had an abstract, the average number of main headings was 5.32, all except four had an acknowledgements section, and the average number of paragraphs of main text was 30.6. The average for figures was 2, that for schemes, 0.4 and that for tables, 0.9.

Art	Ath	Abs	MHd	Ack	Para	Fig	Sch	Tab
HBRO	1	0	0	0	21	1	0	1
HKNU	1	1	6	1	40	0	0	2
HNIC	3	1	5	1	25	11	10	2
HOLS	1	1	6	0	45	0	0	0
ave	1.5	0.8	4.25	0.5	32.8	3	2.5	1.3

Table 10. Summary of history review article components. (Art = article code: H = history; Ath = no. of authors; Abs = abstract present 1, absent 0; MHd = no. of main headings; Ack = acknowledgements, present 1, absent 0; Para = no. of paragraphs of main text (not including abstract); Fig = no. of figures; Sch = no. of schemes; Tab. = no. of tables)

Art	Ath	Abs	MHd	Ack	Para	Fig	Sch	Tab
SCLE	1	1	7	1	32	0	0	0
SGOL	1	1	5	1	20	0	0	1
SLAI	4	1	5	1	11	1	0	1
SLEN	1	1	4	1	27	0	0	0
SRAM	1	1	6	1	59	4	0	0
SSHU	4	1	4	1	31	3	0	3
SVAL	1	1	3	1	23	2	0	0
SVAU	1	1	4	1	23	0	0	0
SWIL	1	1	6	1	43	0	0	1
ave	1.67	1	4.89	1	29.9	1	0	0.7

Table 11. Summary of status quo review article components. (S = status quo. See Table 10 for other explanations)

The overall average number of references for the corpus was 74, with an average of 10.44 (14.11%) being self-citations. The number of references to unpublished data was nil to three in most, with the exception of ITEM, in which there were 19. The expert informant who commented on this review said that it was an extremely specific one which was trying to propose the author's hypothesis about a phenomenon and therefore required supporting evidence. The number of references to personal communication was also nil to two when there was any, with the exception of SWIL in which there were 24. The expert informant comment on this was that the field was a 'hot' one in which it was difficult to keep up to date.

Art	Ath	Abs	MHd	Ack	Para	Fig	Sch	Tab
TAND	1	1	4	1	21	3	0	0
TBEN	2	1	1	0	17	2	0	1
TBRE	1	1	3	0	23	1	0	1
TGRA	1	1	14	1	65	1	0	1
TPRI	3	1	3	1	11	0	0	1
TSCH	4	1	2	1	29	4	0	1
ave	2	1	4.5	0.7	27.7	2	0	0.8

Table 12. Summary of theory review article components.
(T = theory. See Table 10 for other explanations)

Art	Ath	Abs	MHd	Ack	Para	Fig	Sch	Tab
IAME	3	1	4	1	29	3	0	1
IBEU	1	1	9	1	38	1	0	3
IBLE	1	1	4	1	11	1	0	1
IEPS	1	1	7	1	44	0	0	1
IKEY	1	1	10	1	49	0	0	1
ITEM	1	1	11	1	27	3	0	0
ave	1.33	1	7.5	1	33	1	0	1.2

Table 13. Summary of issue review article components.
(I = issue. See Table 10 for other explanations)

An interesting feature is the use of questions as section headings (SCLE, SGOL, SRAM, IKEY, ITEM). One (IKEY) has a question for the title. Quotes are used by two reviews, HBRO (from Homer) and IEPS (from Alfred North Whitehead and Bertrand Russell, *Principia Mathematica*). Also, 'Note added in proof' is found in two of the articles (HOLS, SVAL), indicating the rapidity of developments in these fields.

Art	Ref	Own R	% Own	Unpub	Persl	Pp	Words
HBRO	40	0	0	0	0	3	2237
HKNU	103	10	9.71	0	0	8	7351
HNIC	71	26	36.62	3	0	8	4699
HOLS	80	5	6.25	0	0	7	6140
ave	74	10.25	13.95	0.75	0	7	5106.8

Table 14. Summary of reference and length details for history reviews. (Ref = no. of references; Own R = no. of references including at least one of the authors; % Own = percentage of reference to own work; Unpub = no. of references to unpublished data/results; Persl = no. of references to personal communication; Pp = no. of printed pages (including partial pages); Words = the total number of words in the text itself including the abstract, but not the title or headings)

Art	Ref	Own R	% Own	Unpub	Persl	Pp	Words
SCLE	55	12	21.82	1	0	5	5075
SGOL	49	3	6.12	1	0	5	5299
SLAI	23	7	30.43	3	1	3	2343
SLEN	38	1	2.63	0	0	3	3026
SRAM	57	7	12.28	0	0	8	8390
SSHU	82	16	19.51	0	0	7	5723
SVAL	70	12	17.14	0	4	4	3409
SVAU	103	3	2.91	1	0	4	3351
SWIL	135	1	0.74	3	24	8	7126
ave	68	6.89	10.13	1	3.22	5	4902.3

Table 15. Summary of reference and length details for status quo reviews. (See Table 14 for explanation)

Summarizing these findings presents a profile of the review article (particularly written for the *Proceedings of the National Academy of Sciences of the United States of America*) as usually being authored by a single person and averaging 4,500 words of text in about 30 paragraphs, which amounts to six printed pages with one or two figures/tables. There are almost always an abstract and some acknowledgements made for aid and support. The number of references is quite high, averaging more than 70, considering the length of the paper. Usually some references to the author's own work is included. Almost all use informative headings to set off the sections in the body of the test. Other than the Abstract section, there are no set headings, such as 'Materials and Methods', 'Results' and 'Discussion', which are found in articles reporting current research.

Art	Ref	Own R	% Own	Unpub	Persl	Pp	Words
TAND	68	7	10.29	0	0	5	2824
TBEN	59	11	18.64	0	2	5	3338
TBRE	39	13	33.33	0	0	5	4250
TGRA	76	14	18.42	0	0	8	6232
TPRI	53	10	18.87	0	1	3	1932
TSCH	66	36	54.55	0	0	6	4048
ave	60	15.17	25.21	0	0.5	5	4115.5

Table 16. Summary of reference and length details for theory reviews. (See Table 14 for explanation)

Art	Ref	Own R	% Own	Unpub	Persl	Pp	Words
IAME	217	15	6.91	2	0	8	4989
IBEU	109	29	26.61	0	0	7	4805
IBLE	72	5	6.94	0	0	4	2824
IEPS	129	6	4.65	0	0	7	6686
IKEY	24	0	0.00	0	2	5	5495
ITEM	44	12	27.27	19	1	4	3227
ave	99	11.17	11.26	3.5	0.5	6	4558

Table 17. Summary of reference and length details for issue reviews. (See Table 14 for explanation)

3.8 Classification of reviews based on interviews with specialist informants

The reviews were classified into the four categories after the informant interviews and internal textual analyses of the articles themselves, which will be explained later in this chapter. For the sake of convenience, the informant interview information is presented with the reviews in their classification groups with a brief summary of the information from the informants that led to their classification. More details on the criteria from internal evidence will be presented below.

The specialist informants were given a questionnaire which also asked about the review article(s) they were requested to examine. They were also asked to identify 'moves' or what they thought the writers were trying to do in the texts. Interviews with the specialist informants provided some interesting insights into these articles. One of the most important was that the reviews seemed to be classifiable into four categories: history (H), the historical events or the background of the field is presented; status quo (S), the current position of the field is explained; theory (T), a theory or model is presented in order to synthesize and advance research in the field; and issue (I), some problem in the field is focused on in order to clarify it or draw attention to it.

Here is a summary of responses from the specialist informants with respect to what they considered to be the purpose of the review article and the audience at which it was aimed.

HBRO
Informant: Michell
Purpose of review: "To recall the history of opiate usage, and to bring knowledge of opiate receptors up to date (to 1993)".
Audience: "Broad biological research community".

SCLE
Informant: Lester
Purpose of review: "To inform the academic community of recent developments in this field (the use of chloroplast gene se-

quences to study plant evolution) and to encourage future work. The review critically appraises all current methods, and mentions successful applications of these methods".
Audience: "Aimed at biologists interested in the evolution of plants, especially large scale evolution of families and orders, for which we have virtually no other reliable evidence. This audience includes both those just entering this field or else already investigating but with less than optimal methodologies, but also those following these and other developments in molecular biology even though they have no intention of undertaking such research themselves".

Informant: Newbury
Purpose of review: "To collate the concepts associated with a particular area of research".
Audience: "A sub-set of biologists that is aware of the basic concepts of molecular evolution and plant phylogeny. Because of the level at which the descriptions are pitched".

SLAI
Informant: Strain
Purpose of review: "Draw attention to the growing field in this case of a family of nuclear transcription factors".
Audience: "Basic molecular biologists. PNAS aims at fairly basic fundamental research scientists".

SVAL
Informant: Thomas
Purpose of review: "To review recent work on cloning dynein sub-units to elucidate the structural basis of motor function of dynein. Title, abstract and subheadings make the content clear".
Audience: "Mainly molecular biologists (much discussion of cloning and dissection of the protein into domains). But probably also intended to be accessible to biophysicists etc. interested in cell motility – emphasizes connection between structure and function of dynein".

Informant: Thornhill
Purpose of review: "To provide a summary/overview of the subject area".
Audience: "Expert and professional with invested interest in subject area".

SVAU
Informant: Al-Rubeai
Purpose of review: "To summarize what is now known of the characteristics, significance and regulatory mechanisms of apoptosis. The way it is structured".
Audience: "Researchers specialising in apoptosis and related field. It needs a good background to understand".

Informant: Mitchell
Purpose of review: "To review recent findings in a very rapidly developing field. It is very similar to many other reviews which have been written on this subject recently".
Audience: "A very general audience. The review is very unchallenging and does not reach a challenging conclusion".

SWIL
Informant: Wood
Purpose of review: "To summarise all available data on this topic to bring workers in the field up to date – & to inform those starting out of the possibilities for experimentation".
Audience: "Fairly broad – this is a 'hot' topic at the moment – and all plant virologists are keeping up to date on progress through reviews like this".

TBRE
Informant: Sweet
Purpose of review: "To review transgenic mouse models of lipoprotein transport genes with the aim of stimulating research into making such mice as better models of human lipoprotein disorders and atherosclerosis. This is clearly stated in summary".

Audience: "Directed towards a specific group of scientists and clinicians interested in lipoprotein disorders and atherosclerosis. Very specific topic".

TGRA
Informant: Lester
Purpose of review: "To describe the ways in which flower structure determines pollination; to a wide audience, and to explain these by hypotheses. There is a blow-by-blow description of each flower type in turn, followed by explanations".
Audience: "General biologists interested in plant reproduction ecology, and also researchers in this field. The account is relatively descriptive and non-technical, although it is presented within a framework of problem definition and solution".

IEPS
Informant: Pritchard
Purpose of review: "To point out a gap or deficiency in current dogma, to so stimulate further work in this area".
Audience: "Ph.D. researchers and above".

ITEM
Informant: Wood
Purpose of review: "To propose the author's hypothesis for the explanation of an observed phenomenon, & provide evidence (experimental & evidence from published work) for it. This is a very specific review".
Audience: "Mainly for researchers in this field. It is very specific – not of broad general interest".

The questionnaire responses suggest a gradient of types of reviews "to recall", "to summarize what is now known", "to describe" mechanisms, and "to point out a gap or deficiency in the current dogma". These explanations indicate that there are reviews which are historical overviews, descriptions of the status quo in a field, explanation of the theory, mechanism or model, and pointing out a specific issue, respectively. The responses in relation to the audience indicate a broad audience for historical reviews to more specialized audiences for status quo and theory reviews and a very specific audience for issue reviews. More details from the interviews will support these observations.

3.8.1 History reviews

The first category that became evident from the discussions was the History review, which can offer a summary and synthesis of a field of research, can point out new directions of study or can correct a previous misunderstanding in order to propel the field forward. These reviews are especially helpful for students or researchers who are about to enter a new field of research.

In discussing the HBRO paper, Michell noted that in the 1980's, the field of cell signalling was recognized by ISI (Information for Scientific Information) as the field with the largest body of publications, numbering 2,000 to 4,000 per year. In the 50's and 60's, a Nobel Prize was awarded for work in this field and four Nobel Prizes were presented in related fields in chemistry, physics and medicine, attesting to the importance of the field. Michell said that different kinds of reviews are written at different times in the field. In cell signalling, in the 1970's, there was a large body of literature but little understanding of it all. Therefore, there was a need for someone to summarize and synthesize it into models which could be used to predict, to lead to new experiments, and to move the field forward. He himself had written almost 15 reviews in over 20 years but only three had been synthesis reviews; the rest had been progress reports. He considered the HBRO review to be interesting for the history it covered but that there was nothing new with respect to science. He felt that it was "cursory" and not for "aficionados" of opiate research. He also thought it was well written, "almost an essay" for a nonspecialist audience to understand but not necessarily do research in it. He pointed out that the last paragraph was probably "out of date" because at the time the review was written, there were only two known receptors but at the time of his interview (three years later) at least "half a dozen to ten" were known. He considered it a "good" review because, while it was not expected to "change people's thinking", it could be valuable for young people "to get some feeling for how a field came about".

Another history category review discussed with a specialist informant was HKNU on antioncogenes and human cancer. Lecane had just finished his Ph. D. research in this field on p53, a tumor-suppressor gene involved in about 50% of human cancers. Although he had finished his laboratory work about a year and a half earlier, he found it very difficult to complete the writing because of the explosion of information in this field, which was only three years old. To illustrate this, he immediately did a database search for p53 which yielded 4,497 hits from 1990 to 1996 (as of March 8, 1996). For an even more specific "p53 + human" search in the title, there were 901 hits; there were only 28 and 65 hits in the first two years but 105 in the next, 182 in 1993, 226 in 1994, 243 in 1995, and 52 in only the first two months of 1996. When this search was extended to abstracts and key words, the total jumped to 3,169. He pointed out that the author of HKNU had written a seminal paper on mutation and cancer in 1971, proposing for retinoblastomas the Knudson 1-hit hypothesis, which was general enough to be applicable to other cancers.

The HOLS article was considered by Thomas to be not only a historical re-evaluation of simple reiteration but one that actually went back to the original literature to "put the record straight". He pointed out that the value of one of the references was not truly appreciated until another appeared to make the technique more widely applicable. He pointed out that the lists of examples after the theory explanation was "particularly impressive" but not very evaluative. A standard review article according to Thomas was something in *Annual Review of Biochemistry* or one of the other specialized field journals. He said that if he had to teach how to write a review article, he would cite the need to aim for a scholarly synthesis of everything in the field. Reviews are highly cited and therefore many journals include them. The minireviews have a slightly different aim, being on more topical issues with references to a few recent papers. They would be "interesting and eye-catching" and selective in scope. On the other hand, the *Proceedings of the National Academy of Sciences* aims for as wide an audience as possible. Interestingly, he and a group of about ten wrote a 'review' for a specialized journal which does not "do reviews". They therefore had to adopt the Introduction, Results and Discussion format but did not include any new experimental data in it.

In sum, what became evident from the informant interviews were the following points:

- reviews of different types are written depending on the level of progress in a field;
- after a certain amount of progress, there is a need for someone to synthesize the findings, develop models for application, and suggest future avenues of advancement;
- some reviews are written to re-evaluate the history of a field;
- research and development are progressing extremely rapidly in some fields;
- some reviews are specifically aimed at a nonspecialist or novice audience ['nonspecialist' refers to someone within the scientific discourse community but not in the specific field].

Thus, history reviews can be useful for a nonspecialist or novice to quickly learn about a field, its models, and avenues of possible research. They can also help amend the history of a field.

3.8.2 Status quo reviews

The Status quo reviews were found to be aimed more at the researcher working in a closely related field to keep aware of updates and recent literature, especially in rapidly developing fields of research. When asked to examine the review articles, several of the informants searched for 'well-known names' in the field. It seems that these review articles can offer an efficient way to keep up with major trends in the field.

The SCLE paper was presented to Lester, who immediately checked the references for a seminal paper in this area, which changed the entire field of plant classification, and was satisfied when he found it. He said the information in the review article under discussion was important, but too complicated to introduce to students, who would not yet have the relevant background knowledge, although as it contained revolutionary material, he would tell the students about it. He said the field of chloroplast gene sequencing itself had only begun from about five years earlier. There was an explosion of progress after free samples for analysis were distributed by one of the researchers.

For the SVAU review, Mitchell sent a note saying that doing the linguistic analysis had been a "very difficult exercise". He went on to state

> This particular review may not be a good paper to work from. As this field underwent very rapid expansion, hundreds of similar reviews were written. Few, including this, have anything original to say [...] more just a repetition of what everybody in the field knows is in the literature. It's rather like a very clearly written undergraduate essay.

The field of physiological cell death essentially did not exist four years ago (interview 1996), and when one of the references quoted originally appeared, saying that cells could commit suicide, the author was treated as an outcast. Then in the early 1990's, the field exploded and he himself became involved in 1993-94 and is interested in its relation to neurobiology of the brain.

The SLAI article was reviewed by Senior Lecturer Alistair Strain of the Liver Research Unit who stated that although he thought the *Proceedings* to be "a good journal of high scientific standard," he would go to more narrow state-of-the-art journals such as *Cell*, which has minireviews, or *Journal of Cell Biology* or the *American Journal of Pathology*. He uses reviews for updates on recent developments, access to recent papers for specifics to read, for names or addresses for contacts for access to reagents, and most importantly, for update of literature coverage. He was rather critical of the SLAI article as his specialty is the liver and he thought that the "biological world in general thinks that the liver is a good place to do research on cell- or tissue-specific factors" but he thinks otherwise. He also said that the one of the authors listed "tends to simplify things, but this is what molecular biologists tend to do". He said that nuclear transcription factors had been known for 20 or more years but that the details and understanding were progressing rapidly only in the past three to five years.

The SWIL article is unusual because it has 24 instances of reference to personal communication. Newbury's first response was that it was "strange" then after skimming through the article, said that it was probably because the author had recently moved to a different laboratory and was probably "getting the latest data from his colleagues". He said that in this field, it takes about 9 months to publication so that the latest data would not be "in print" for about 6 months. The field covered by the review had begun about 10 years prior to its publication but there are many central concepts which still need to be discussed because there are many different crop plants and many different viruses.

This SWIL review was also discussed with Wood, who contrasted it with the ITEM one which he considered to be unusual. He said that the SWIL was a typical review which surveys recent literature and comments on significant things. He considered it to be a review that would be useful to all kinds of people, both teachers and researchers in the field. It is of broad interest and a good summary with repetition of similar moves throughout. Wood, like some of the others, mentioned that he would usually read journals that were more field-specific and would use a database to search for subjects or authors of interest. He would not usually peruse through journals. The author of the SWIL review is well known in the field and "one of the names he would look for". If he were to teach how to write a review, this is one he would use to provide pattern guidelines.

To summarize, the informant interviews revealed that:

- status quo reviews are usually aimed at practicing professionals who use them to keep up-to-date in not only their own fields but also related fields;
- author name value and the suitable citing of papers is very important;
- as mentioned by the History review informants, progress is extremely rapid in some fields;
- specialists in different fields view findings in different ways;
- there seems to be an extent of freedom of expression as one of the articles had an unusual number of personal communication references which were suggested to be due to an attempt to introduce concepts which were not yet published, perhaps in an attempt to circumvent the time-consuming publication process.

The status quo reviews are useful for researchers and teachers in the field or in related fields to keep abreast of developments.

3.8.3 Theory reviews

The third category also seems to be a review written more for researchers in the field. The Theory review, while not offering new information, seems to be used to introduce a theory or a model(s) to further work in a field.

Lester, who discussed the SCLE paper mentioned above, also commented on the TGRA paper. He noted that the author was an excellent person to write such a review. The writer was a member of the Department of Botany, University of Texas, which had been voted by peers in the United States as the best department of botany in the United States. Lester himself had studied there in the 1960's. As for the content, he felt that there was nothing really "new", meaning that he found no new data or evidence, and that the article was only trying to present a "new model" for well-known facts.

The TBRE article was considered by Sweet to really be "several introductions" composed of a general one and subsections with others. He thought that the author was trying to convince others in the field that the mouse model with adaptation could be used for atherosclerosis because it can be used for lipoprotein diseases. This review was not aimed at the general audience but for people in the field. He considered the review article to be good as a teaching instrument. The two main reasons for this are that they serve as a summary of one's own work for presentation to a wider audience and they can draw together things in a field one is not too familiar with to be useful for others. His laboratory had been involved in the writing of two review articles, the first a summary of their own work and the other on how viruses cause disease. The latter was occasionally cited but as a simple referral as an information source. He found it useful to read reviews when entering a new field and did not have enough time to read all the references. He also finds them useful for teaching.

In sum, these informant interviews on Theory reviews showed:

- the importance of the author's qualifications to write a review;
- their convenience as quick information reference sources;
- their utility as teaching materials.

The Theory review, unlike the History review, focuses on developing or supporting a specific theory or model and can be useful as a teaching material. It can also be used by writers to present their ideas to a wider audience than a research article which would have a more narrow and specific focus.

3.8.4 Issue reviews

The features of the fourth category of review articles mainly came from analyses of the texts themselves because only one of the articles could be discussed with a specialist informant. The ITEM review was commented on by Wood who considered it to be very different from SWIL, which he had also examined. Wood felt that the author was trying to propose a theory and covered only a very limited, specific area. He thought that the experimental data presented was "unusual" and that while the article was "reviewing phenomena", it was also "introducing the author's own views and data". He said that the author was one of the discoverers of the retrovirus and one of the world's leading specialists. Wood thought that the writer was trying to propose a mechanism for virus replication and that it was difficult to understand and evaluate if one were not in this specific field. Wood felt that writing reviews could be "an ego trip for the author" while he considered that the "primary reason should be to inform the audience".

From the textual information shown in the following examples, the reviews in this category are considered to focus on some specific issue(s) in their respective fields of study. The ITEM author clearly states: "I propose that the numerous mistakes in reverse transcription are the result of [...]." The other Issue reviews also use phrases such as [...] *has created a host of [...] dilemmas, omission of [...]. may lead to distorted results, opportunity for improving health by improving diet is great.*

3.9 Type classification of reviews based on text evidence

Informed by the discourse community specialists, I next tried to classify the reviews based on textual information. For the history category, phrases such as *a brief history of [...], review progress,* and *summa-*

rizes the main contributions to the field were taken as indicating a historical review. The status quo review judgment was based on the use of phrases such as *a brief account of the history and status of, summarizes the current state of knowledge,* and *summarize the recent [...] information*. The theory reviews presented discussions of models which the authors wished to promote to aid further progress in the field and were identified by the use of phrases such as: *this information is used to construct several models, propose a model, provides a perspective on*. The issue category was one in which the author(s) was raising questions or issues about something in the field and was signaled by phrases such as: *a paradigm of current issues, omission of [...] may lead to distorted results, I propose that the numerous mistakes in [...] are the result of [...]*. In sum, both the specialist informant information and consideration of the textual information led to the following four categories:

1) History: presenting historical view of a facet of the field;
2) Status quo: describing the current situation in a field;
3) Theory/Model: proposing a theory or model to resolve some issue in the field;
4) Issue: calling attention to some issue in the field.

The History reviews tended to have words and phrases suggesting a summary of the history or progress in a field. The Status quo reviews used words related to how recent the information is: *current state, news, recent.* The Theory reviews had many references to *models* and *analysis* and other words related to classifying information, such as *code, mode* and *role*. The Issue reviews had words suggesting problems that needed to be addressed: *anomaly, dilemma, distorted results, issue, limit, mistake, problem* and *risk*.

This chapter has shown that the review article can be considered to be a genre from both the specialist informant viewpoint and bibliometric data. The review article seems to be a means by which researchers can summarize, evaluate, synthesize and make projections with ideas in a field of research for not only those in their discourse community but also those in other communities.

4. Microanalyses of review articles

> The revised manuscript by [...] is very much improved. The authors have responded to all of the previous criticisms. This paper is I believe now acceptable for publication in PNAS. However, the manuscript still has numerous errors in English grammar and sentence structure, particularly the figure legends. I would suggest, therefore, that the paper be carefully edited by someone with expertise in English writing before the paper is sent to the Proceedings of the National Academy of Sciences for publication.
>
> (anonymous reviewer, January 27, 1991)

4.1 Introduction

The macro-level considerations of Chapter 3 offer general information frameworks but more detailed micro-level analyses are needed at the sentence, clause, and phrase levels in order to be of aid to a non-native English speaker as can be surmised from the reviewer quote presented above. This chapter will describe a way of analyzing the review texts to enable identification of micro level features which should be helpful for developing writing support tools for the NNS scientist. The overall analysis framework to be used is presented in Tables 18, 19 and 20. Micro-level features of Discourse Steps, Moves and Move Steps are examined.

Physical feature	Move	Steps
Rhetorical organization: Beginning		
Title		
Affiliation		
Abstract		
History	Present topic of paper	Topic
	State what is known	His, Phe
	Point out problems (optional)	Tech, Gap
	State aims of review (optional)	Aim, Disc
Status quo	Present topic of paper, usually with emphasis on recent work	Topic
	Describe recent research/findings and evaluate them	CR, Def, Phe, Tech, Ela
	Give opinion about recent work	AsFL, Op
	State aims of review article	Aim
Issue	Present topic of paper, sometimes including indication of the problem to be discussed	Topic, Phe, Def, Imp, Gap
	Describe known information	Phe, His, AsFL, Ela
	State opinion, point out issue	Op, Hypo, Gap
	Present hypothesis/opinion about a single issue	Hypo, Op
Theory/Model	Present topic of paper	Topic, Def, His, Tech
	Describe what is known	Phe, Def, CR, Ela, Tech
	Offer overall theory or model or explanation	Aim, Hypo, Dev
Quote		
History, Issue	Attract interest	

Table 18. Framework for review start.

Physical feature	Move	Steps
Starting paragraph(s)		
History	Present topic	Topic
	Present historical background, usually chronologically	HisH, Phe, Ela, LitH, CR
	Point out any gaps in knowledge	Gap, Hypo, Op
	State aim of article (usually to review what is known)	Aim
Status quo	Present topic (often with a heading and reference to historical information)	Head, Topic, His, Phe, Def, CR, Tech, Aim, Op, Disc
	Present what is known	Phe, Def, CR, Ela
	State aim of article (usually to present current status of field; key word: recent)	Aim
Issue	Present topic	Topic, Imp, Phe, His, Op
	Point to the existence of an issue/problem	Hypo, Op, Gap
	State aim of article (usually to give opinion on an issue)	Aim, Op
Theory/Model	Present topic, give definition	Topic
	Describe phenomena	Phe
	Present history and current research	His, CR
	Point out need to synthesize current knowledge	Gap, Hypo, Op, Dev
	State aim of article (usually to present own theory, model or explanation)	Aim, Disc, CR

Table 19. Framework for review initial paragraphs.

Physical feature	Move	Steps
Rhetorical organization: Middle		
	Use section headings to guide reader	
Rhetorical organization: End Summary / Conclusion		
History	Indicate closure	Head
	Review history, current research level	His, CR
	Offer predictions about future work/trends	Hypo
	Point out remaining gaps in knowledge	Gap
	Present opinion as closing statement	Op
Status quo	Indicate closure	Head
	Summarize	Sum, Op
	Note gaps, hypotheses	Gap, Hypo, AsFL
	Suggest future work	CR, Dev
	Present opinion as closing statement	Op
Issue	Indicate closure	Head, Sum
	Restate topic	Topic
	Suggest explanation, future trends	Hypo
	Present opinion on issue as a closing statement	Op
Theory/Model	Indicate closure	Head
	Summarize proposed explanation	Sum, Phe, Hypo, Op, Tech, Gap
	Present opinion about future applications of theory, model, explanation to serve as closing statement	Op, Hypo

Table 20. Framework for review middle and ending paragraphs.

After the main text of the review, other physical features observed were ‘Note in Proof’, ‘Acknowledgements’ and ‘References’.

4.2 Moves and Steps

At the start of this research, an attempt was made to use move analysis arising from the seminal work of John Swales in genre analysis (1981, 1990a). Swales proposed that the introductions of science research journal articles were trying to do three things: establish a territory, establish a niche, and occupy the niche. In other words, the writer would try to convince the reader of the importance of the research field, point out a problem or gap in the field knowledge, and present his/her research to fill that gap. As reviewed by Dudley-Evans and St. John (1998), many others have followed this approach to analyzing the functions of different parts of a text. Move analysis has been applied to MSc dissertations (Hopkins and Dudley-Evans 1988), job application letters (Bhatia 1993) and the discussion section of medical research articles (Hozayen 1994). These concepts have also been incorporated into pedagogical materials for teaching the writing of the research article by Weissberg and Buker (1990) and Swales and Feak (1994).

With this substantial work in move analysis, the task of finding how to help NNS with the writing of the review article seemed as though it would be a rather straightforward task, involving analysis of the moves used in a review article followed by a cataloguing of the words and phrases that tended to appear with high frequency in specific moves. Such findings could then be used to prepare an index of words and phrases frequently used in such moves to help the nonnative speaker (NSS) writer.

However, most of the previous move analysis work had been with limited portions of text with more specifically identifiable aims, unlike the review article which, as several of the informants commented, are like essays. One of the analyses attempted was to have the specialist informants match the moves from a list drawn up based on my initial text analyses. A letter was sent to the specialist informant together with the review article printed out with page and line numbers. The letter requested permission to tape record the interview session but most of the informants indicated that they would not feel comfortable with this, and thus, only detailed notes were taken at each session. The moves initially listed in the letter were:

List of possible author moves in text

1. Expressing writer *aim*
2. Describing the *importance* of the field
3. Describing a *gap* in the knowledge of the field
4. Giving a *definition*
5. Introducing a *topic*
6. Describing generally accepted *phenomena*
7. Introducing *literature*
8. Introducing *supportive literature*
9. Trying to *explain phenomena*
10. *Assessing* findings, literature
11. *Summarizing*
12. Explaining *limitations*
13. Giving *opinion* – judgements as to value, validity
14. Hypothesizing about phenomena – predicting about how phenomena work
15. Explaining possibilities for *further development* and applications
16. Making a *recommendation*
17. Organizing *discourse* – referring to discourse elements

The informants were told that the list was only a suggestion and that they could use other classifications to describe what they thought were the writer's intentions. Based on feedback from the first informants contacted, the list was expanded to 22 items with the following additions:

18. Describing key *technologies*
19. Describing goals of *current research*
20. Evaluating *history*
21. Explaining underlying *theory*
22. Presenting *examples*

The results of the move analyses by the specialist informants were not very encouraging. The informants found the move analysis more difficult than expected and none were able to complete the analysis as requested. For those who did attempt an analysis, some wrote their own interpretations in the margins or on a separate sheet: "description of a sixth component", "restatement of hypothesis" (Wood for ITEM), "attempt to explain the phenomenon", "assessment of published data" (Wood for SWIL), "example 3, assessment/opinion/conclusion" (Sweet for TBRE). Such results or the lack of response (in some cases) and comments such as the following:

> I have tried quite hard to do something useful for you but I am far from convinced that I have. At the end of the day it's a very difficult exercise [...] I hope some of this is of some use. (Ian Mitchell, University of Birmingham, March 1, 1996)

indicated the difficulty of move analysis even for native English speakers who are experts in their fields and many of whom have written review articles themselves. In other words, if the moves could not be identified by native speakers of English, then they probably would not be usable for NNS writers.

Another problem encountered with move analysis was the fact that it was difficult to identify the boundaries of a move. For example, some of the responses were comments such as "page 3, 15-29, describing process of cell death which is now a generally accepted phenomenon" and this was immediately followed by "35, clarification of term 'programmed cell death' which is ambiguous". Lines 30 to 34 were left unmarked.

In order to have an analysis framework that could identify features helpful for the development of NNS tools, there was a need to work at a linguistic level below that of the move and a decision was made to examine all text levels starting from the word/phrase and the clause to the sentence, paragraph and section in an attempt to identify useful, usable features. Swales (1990a: 142) also found it necessary to break down the moves into parts that he identified as 'steps'. This approach at a level below that of the Move is also discussed in the work of Sinclair and Coulthard (1992: 4), who stated:

> We then realized that moves too can have a structure and so we needed another rank with which we could describe this structure. This we labelled *Act*.

They were working with the analysis of spoken discourse, in this case with classroom discourse; however, they essentially were analyzing transcriptions of the discourse and admit that "this research has been very much text-based" (Sinclair and Coulthard 1992: 4-5). For example, in describing discourse acts, they state that these are "typically one free clause, plus any subordinate clauses [...]". In this book, the following analyses are of Steps mainly based on the orthographic sentence, except where a clause (such as when a compound sentence has two distinct parts) or a phrase (such as a section heading) can be specifically identified as an independent step. In searching for a way to satisfactorily analyze the moves of review articles to discover insights for NNS tools, this concept of steps, or acts, sometimes clustering to form moves, was adopted for the present research as it was expected to be useful in actual practice.

4.3 Move and discourse organization

In this work, a Step is defined as the action mainly expressed by an orthographic sentence, but could sometimes also be expressed by a phrase or a clause, as discussed above. These Steps were the 'moves' which were presented to the specialist informants when they were first asked to classify the textual information. Based on their feedback and the following features, the Steps were used as the basic unit of the genre analysis in this book. These Steps are subjected to further analyses in Chapter 6. The features used to classify a Step include the sentence initial element (SIE), grammatical subject (GS) and verb tense and modality (Vb), which are explained below.

The Discourse Organizer Step is applied to sentences/clauses/phrases which are predominantly functioning to organize the text or present discoursal signals to the reader. These seem to differ from the Steps which are mainly related to the textual information itself and therefore have been placed in a separate category of Move Steps (explained below). These Discourse Organizer Steps are global items used to organize the review article at the whole text, section and paragraph levels. They, i.e., Topic, Aim, DiscourseSignal, and Sum, primarily serve as signals to guide the reader through the text but can often be assigned to a clause together with a Move Step. As both Discourse Organizer Steps and Move Steps are composed of similar features, the respective Steps will be described together in detail below.

4.4 Step features

One of the three major aims of this book is the development of useful writing support tools for nonnative English speaker writers. In order to do this, the detailed features of each step will be examined next: SIE, sentence initial element; GS, grammatical subject; and Vb, verb tense or modality. These features will then be used to discuss the Step types. The relationship between Steps, Moves and review type will also be explored.

4.4.1 SIE – sentence-initial element

As the sentence is considered the basic unit of the Step, one of the obvious features to examine is how the sentence is started. Also, as discussed in Chapter 2, among the nested issues in NNS writing are problems with the appropriate use of discourse signaling. Such signals are often found at the beginning of sentences, paragraphs or sections. Therefore, one of the Step features identified for all sentences in the corpus was the element used to begin it. Here I would like to borrow the framework proposed by Gosden (1992a) for these sentence-initial elements (SIEs). Gosden's work follows that of Davies (1988), who is cited as pointing out the need to make evident the words and phrases as well as the grammatical items that make possible the expression of the different rhetorical functions in a research article. As discussed in Chapter 2, this is also one of the basic concepts of this book, that there are nested issues to be resolved in order to aid the NNS writer. Gosden (1992a: 221) suggests that analyzing for marked theme could reveal genre features which would enable NNS to gain "greater control of the writing (and rewriting) process, thereby helping to raise students' learning ego".

As Gosden's aim was to "investigate the potential that analysis of marked theme has for understanding the tendencies and propensities of within-text structuring of scientific RAs (research articles)", he focused on only the marked themes and categorized the sentence-initial elements which function as a "broad range of contextualising frames (CF)" (Gosden 1992a). He identified the following three grammatical categories:

(1) Conjunctive/modal adjuncts and conjunctions – minimals: *briefly, however, therefore*; modal adjuncts: *certainly, evidently*; conjunctions: *but, and, yet*

(2) Prepositional and adverbial phrases – simple prepositional phrases: *in this paper, among these techniques*; complex prepositional and adverbial phrases: *because of the similarities, in the case of x = 0.92, in the light of these results*

(3) Subordinate clauses and nonfinite clauses – *while the fraction was x, when fk = 1 is applied, referring to the data of Smith, unless otherwise stated, as shown in Figure 1*

Gosden (1992a) next identified the functional categories shown below. Where applicable, he presented examples for each of the grammatical categories given above. As described above, the work in this book is also aimed at finding ways to aid NNS writing but the unmarked themes and how they are used also appeared to be of importance, as will be seen in the discussion of some of the Steps. Therefore, in the following list, I have included other sentence-initial elements (SIEs) which are not necessarily marked themes. They should help identify what is at the beginning of the sentence (or clause) in order to establish Step features. Given below are the SIEs (Sentence-Initial Elements) from Gosden (1992). [*indicates a category that I added to help identify Step features].

AddA Addition appositive
for example, for instance, in other words, that is

AddE Addition emphatic
furthermore, moreover, indeed, besides X,
in addition to Y, besides being effective, taken together

CauseP Cause, Purpose
to this end, for evaluation of the gap, in order to obtain X

CauseR Cause, Result/reason
thus, therefore, consequently, hence, as a result, due to X, because of Y, since there is Z

Comp* Comparison
similarly, like, coincident with

ComV*	Command form verb *note, see*
Con	Contrast/concession *however, by contrast, on the other hand, instead, despite this information, unlike the present data, although there is controversy, while X was 59%*
CondH	Condition, hypothetical *in principle, perhaps, conceivably, with such possibilities, assuming that V is constant, if X works as Y*
CondR	Condition, real *in this case, then, under these conditions, in the case of x = 0.81, if the polarity is reversed, unless otherwise stated*
It	It/There at start of sentence*
Means	Means *in this way, by further studies, with this construction, using X, by using Y's law*
Ques*	Question verb, *wh-*
S*	Subject is Sentence Initial Element
SpDE	Location in space, discourse entity *here, in this paper, in Figure 1b*
SpRWE	Location in space, real world entity *here, there, on a glass substrate, below the spectra, where uptake occurred*
TRWE	Time Real world entity *recently, previously, subsequently, later, after the removal of X, as the temperature increases*
TDE	Time, discourse entity *in summary, in brief, initially, throughout the discussion*
Val E	Validation, external *in general, usually, normally, in previous studies, from the data of Smith, as is well known, as noted earlier*
Val I	Validation, internal *significantly, clearly, evidently, from the results in Figure 1, As shown in Fig. 2*
View	Viewpoint *interestingly, unfortunately, surprisingly, in the broader view*

4.4.2 GS – grammatical subject

Another feature examined for the Steps is GS which refers to the Grammatical Subject and the categories of the subject. Again, the work of Gosden (1993) is used as a basis for the analysis with some of the modifications proposed by McKenna (1997). Gosden points out that "one key element for perceived success in written communication is a writer's ability to judge the appropriate balance between interactional 'human face' discourse and impersonal topic-based discourse, the 'dynamics of context building' being determined by the requirements of different written tasks and genres" (Gosden 1993: 56). This led him to try to model the subject role domains in research articles for degree of writer visibility and community orientation. Summarized below are the categories of discourse functions of the grammatical subject of the clause, with material from Gosden (1993a) [G] and McKenna (1997) [M] with some of my additions and interpretations [J].

Participant

PIntP	Interactive participant
G	Researchers referred to by name in citations *Smith 1987, none of those authors*
M	Direct reference to the person or corporate entity that initiated the report
Ppart	Participant viewpoint
G	Focus on research outcomes/activities; our research rather than our reporting *Our argument, our data* (not used by M)
M	Any other participant viewpoint using *our*
PDisP	Discourse participant
GM	We, I

Discourse

Dint	Interactive discourse entity
GM	Referenced units which refer to previous community-validated discourse entities *Previous studies, little is known; there have been few published studies, there has been great interest, many papers have appeared, a number of parallel studies*

Demp	Empty discourse theme
GM	Frequently introduce evaluative comment and therefore play a significant role in interactional thermatization *It is concluded, it can be argued; there are reports, there have been reports*
Dmicro	Micro-discourse entity
GM	Part, discourse-internal entities *Fig. 1b, diagram, connected reference items*
Dmacro	Macro discourse entity
GM	Integral units of discourse *This paper, report, communication* *thesis, work*
DDE	Discourse event/process
G	
M	Discourse Steps and processes; frequently nominalized reporting verbs *The conclusion, interpretation, argument, explanation, observation, description*

Hypothesized/Objectivized

HOAn	Analytical concepts of real-world entities, processes and events
M	Scientific naming as 'a linguistic construal of experience' (Halliday and Martin 1993; 7-8) *Major cleaners include; Solid wastes consist of;* *drag-coefficient* (named items)
HOHe	Hypothesized entity
GM	Modes of testing and carrying out research *The model, approaches, formulae*
HOEmp	Empty H and O theme
GM	*It was evident, it is interesting to note* *There was evidence, there are two possible contributions*
HOOv	Objectivized viewpoint
GM	G: High degree of evaluative modification implies acknowledged or given status frequently reinforced by use of timeless present tense; lacks modal *a significant difference, the most striking influence; One factor, a further contribution, further evidence; The most*

	striking influence of the P-phase; What is particularly striking
J	*utility* (BRO 9.5)
OIC	Objectivized intellectual corpus
M	The name of general relevant objectivized information; only one word now: *data*
J	*knowledge* (BRO 4.2)
HOHv	Hypothesized viewpoint
G	
M	Comment and judgement on research matters
	The possibility, the most probable cause, the apparent contradiction, one possibility

Real-world

RWT	Empty real-world theme
G	Postponed real world entities, researchevents /processes and reference to mental processes
	It was found; There was a delay, there is a small further red shift
RWE	Real-world entity
G	Material entities and objects concerned with the physical world
	AlFeNi alloy
J	People mentioned as entities (BRO 3.3)
	Actions and procedures **not** executed in or resulting from scientific activities (BRO5.5)
RWP	Real-world event/process
GM	Actions and procedures executed in or resulting from scientific activities
	Annealing, the preparation of the amorphous samples
M	Events or processes under the engineer's consideration; entities under consideration
RWM	Real-world mental process
GM	Focus on intellectual processes and entities
	Deduction, calculation, comparison, analysis, determination, evaluation, estimate, assumption, confirmation, idea, insight, question, prediction

RWat Non-mental engineering analytical tool
M Entities used in the investigative process
photographic record, monitoring; a great deal of effort

Other
Oth Other, e.g. demonstratives and imperatives
M Demonstratives are allocated this category
Imperatives also
This leads to buckling of [...]
These vibrations – real world entity subject as an unmarked theme
J *None* (BRO 7.5)
Non Non-sentence
M Sentence without a main clause and/or finite verb

4.4.3 Vb – verb tense/modality

Vb refers to the verb tense or modality of the main verb in the sentence or clause. This was a distinguishing feature of a Step as the present tense was often used when describing knowledge that was considered to be accepted as fact by the discourse community, while the past tense was frequently used when referring to a single report of a study. As can be seen from Table 21, the present perfect tense was often used when referring to work by multiple investigators or developments which remain ongoing.

Verb tense	History %	Status quo %	Issue %	Theory %
Present	50.69	44.90	59.14	60.73
Past (including past perfect and continuous)	23.79	20.93	13.86	10.10
Modal	12.39	19.66	17.91	17.40
Present perfect	12.77	13.77	8.82	11.46
Present continuous	0.38	0.49	0.27	0.10
Imperative	0.00	0.18	0.00	0.21
Future	0.00	0.06	0.00	0.00
Total no.	799	1648	1111	960

Table 21. Verb tense/modality in the four categories of review articles.

4.5 Step types

This section will attempt to identify specific features of the Steps to enable their use in teaching English as a foreign language. In order to do this, the Step types presented in the list given below were identified by first considering what the author was trying to do in a review article then asking specialist informants to try to use them to identify author intentions in the text. Their feedback was then used to further refine the definitions together with the linguistic features described below.

The Steps were assigned before trying to analyze their features in order to prevent a Step being assigned merely because of its surface features, which is a problem pointed out by Dudley-Evans and St. John (1998) about becoming 'stuck' in the text. However, after the assignments had been completed, some had to be revised because discrepancies were sometimes noted between the earlier assignments and the later ones. The rationale for this was that Steps would embody both form and function and therefore their analyses would probably be a circular process taking into consideration both the linguistic features and the intuitive intent. As a heuristic approach, the clauses were first tentatively assigned to various Steps then the following features were examined to try to determine the individual linguistic features in order to link them to the review type.

The Steps used for the final analysis could be classified into four types which organized, presented, evaluated or synthesized information. The organizing Steps were considered to be Discourse Steps which functioned at a level different from the Move Steps, which were closely related to the content information itself. These Discourse Organizer and Move Steps will be analyzed separately in the next chapter.

Discourse Organizer Steps: Organizing

Topic	Introducing a topic
Aim	Expressing aim
Disc	Organizing discourse – referring to discourse elements, using rhetorical questions, imperative verb
Sum	Summarizing

Move Steps: Presenting

Def	Giving a definition
His	Describing historical events
Phe	Describing generally accepted phenomena
Tech	Describing key technology, method, tool
CR	Describing current research – not specific, general and current
Ela	Presenting examples, explanation
Lit	Introducing related literature
LitH	Introducing historical literature

Move Steps: Evaluating

HisH	Hypothesizing about historical events
AsFL	Assessing findings, literature
Gap	Describing a gap or limitation in the knowledge of the field

Move Steps: Synthesizing

Hypo	Hypothesizing – about specific phenomena: trying to explain
Dev	Explaining possibilities for further development and applications
Op	Giving opinion – expression of judgments of value, validity

4.6 SIE features and Step types

The tables in this section present quantitative data for the Steps with respect to SIE, Vb, and GS and also the percentages of Steps in the different articles. Examples of frequently found SIEs for specific Step types are presented below in this section.

	Discourse Organizer Steps			
	Topic	Aim	Disc	Sum
Total No.	240	28	130	56
GS	**74.58**	**50**	**42.31**	**35.71**
Con	2.92			
TRWE				14.29
SpDE			9.23	
CauseR				21.43
ComV			6.92	
Ques			20.77	

Table 22. SIE (sentence initial element) in Discourse Organizer Steps (%). (All Steps from all articles were combined into a single file and the percentages of the different SIEs for each Step were calculated. The most frequently used SIE for each Step is given in bold. Very infrequently used SIE types are not shown)

4.6.1 SIE: Grammatical subject (GS)

As can be seen from Tables 22, 23 and 24, the most common SIE for almost all Steps is the grammatical subject of the clause. This is particularly high for the Define Step, examples of which are presented below. For these and other examples presented in subsequent sections, the four capital letters at the head of each example give the review article from which it originated and the numerals represent the paragraph and sentence number, respectively.

Here are some examples of Grammatical Subjects from different types of reviews:

HKNU 3:4 *Such genes* are known as antioncogenes, or tumor suppressor genes.

IBEU 4:1 *Gaucher disease* is one of a group of lysosomal storage disorders.

SWIL 4:6 *TSWV* is a serious pathogen with a very broad host range and is the type species of the new tospo virus genus in the arthropod-borne family Bunyaviridae (23).

TSCH 6:2 *These minimalist synthetase-minimalist RNA substrate interactions* are the biochemical basis of an operationally defined RNA code.

	Presenting Steps							
	Def	His	Phe	Tech	CR	Ela	Lit	LitH
Total No.	68	252	1122	232	196	105	588	52
GS	**94.12**	**50.40**	**65.06**	**59.05**	**56.63**	**51.43**	**58.67**	**61.54**
CondH			8.65					
CondR		5.16		6.90		8.57	4.93	
Con		5.95	3.47	3.02	4.59		4.08	
TRWE		10.71	0.80	3.02			4.25	
SpRWE		2.78	2.94	3.45	4.59		4.25	
SpDE		1.19	0.71					
Means							1.70	
It							1.70	
AddA			1.69			10.48	1.87	
AddE		4.37	2.32				2.38	
CauseR		3.97	2.05			8.57		
Comp			0.53					
There			1.07					
View			1.42					

Table 23. SIE (sentence initial element) in Presenting Steps (%).
(See Table 22 for further explanation)

Lower percentages of the Grammatical Subject were observed for the SIE of the following Steps: Sum (35.71 %), DiscourseSignal (42.31 %), Gap (42.86 %), and History (50.40 %), which had relatively higher percentages of other SIEs, such as CauseResult, Question, Contrast/Concession, and Time Real world entity, respectively.

	Evaluating Steps			Synthesizing Steps		
	HisH	AsFL	Gap	Hypo	Dev	Op
Total No.	15	110	231	576	12	509
GS		**51.82**	**42.86**	**53.82**		**53.44**
CondH				6.77		
CondR			3.46	5.73		5.11
Con		6.36	14.72			5.70
It						6.09
AddE						2.75
CauseR			3.46			5.11
There			3.46			
View			4.33	2.60		7.27

Table 24. SIE (sentence initial element) in Evaluating and Synthesizing Steps (%). (See Table 22 for further explanation)

4.6.2 SIE: Question

An interesting feature is the use of questions as discourse signals for the DiscourseSignal Step. This was most frequently found with IKEY (12 examples) and SRAM (9 examples) but was also found in other articles.

IKEY 36:2 *In the play of interests that constitutes democratic politics, how can such fine tuning of prices emerge?*

SRAM 48:1 *How general are the findings we have reported here?*

HNIC 15:4 *Could this putative receptor serve as a capturing and delivery system for these enediynes to specific sequences of DNA?*

TGRA 47:3 *What are the later-generation effects of the hybridization?*

4.6.3 SIE: Location in space, discourse entity (SpDE)

A relatively frequently occurring SIE is SpDE, which points to a different section in the same paper, such as the following:

HKNU 9:11 *As we shall see later,* the apparent explanation is that for these other tumors, other controls on growth must be overcome; RBI is not the only gene that must mutate.

IEPS 21:4	*In this section on growth,* however, some experiments with soil-grown plants will be included [...].
SLAI 2:2	*Shown in Table 1* are representative members of this rapidly growing gene family.

4.6.4 SIE: CauseResult (CauseR)

CauseR SIE (21.43%, 10 of the 11 are *thus*) was found in a higher proportion for the Sum Step than other Steps. Some examples are as follows:

TSCH 21:3 Rx	*Thus,* >400 amino acids from the nonconserved domain make no contribution, even indirectly, to acceptor helix aminoacylation (48).
HKNU 18:8 xx	*Thus,* there is specificity not only for an organ but also for the degree of differentiation within that organ.
IBEU 31:5 x	*As a consequence,* there is relatively little relationship between the severity of disease and whether or not a patient receives treatment.

4.6.5 SIE Contrast/concession (Con)

The Con SIE registered a relatively high frequency of 14.72% for the Gap Step, such as in the following sentences.

HKNU 18:4	*However,* familial neuroblastoma is rare and linkage analysis has not been conducted.
IBEU 0:5	*Although* it is clear that these mutations provide a selective advantage in the Jewish population and thus constitute a balanced polymorphism, the nature of the advantage is unknown.
SWIL 2:1	*Despite* extensive and sometimes elegant experimentation, the molecular mechanism(s) of this viral "cross-protection" have remained elusive and controversial.
TBEN 13:9	*In spite of* extensive screening, no synaptophysin homologues have yet been identified in other systems.

4.6.6 *SIE: Time Real world entity (TRWE)*

This SIE was found in 10.71% of the History Steps where the time explanation was given as a framework. Some examples are:

SCLE 9:2 — *In 1967,* Sarich and Wilson (20) used a molecular clock argument to posit that the human-chimpanzee-gorilla separation was much more recent (~5 million years) than was believed in the mid 1960s.

TBEN 1:6 — *In recent years,* a combination of approaches, [...], has led to the identification of a number of proteins that are involved in membrane trafficking.

4.6.7 *Summary of SIE findings*

As a heuristic tool, the SIE should be valuable, especially for nonnative English speakers. An awareness of the SIEs usually used for certain Steps should facilitate discourse organization and expression. How this information can be used will be discussed in more detail below.

4.7 Verb tense/modality features of Step types

Presented in the following tables are details on verb tense and modality usage in the review articles examined.

	Discourse Organizer Steps			
Step	Topic	Aim	Disc	Sum
Total No.	240	28	130	56
Past	7.92		2.31	7.14
Pres	**57.50**	**75**	**69.23**	**57.14**
PresP	24.17		6.92	16.07
M	8.75	25	16.92	19.64

Table 25. Verb tense/modality in Discourse Organizer Steps (%). (All Steps from all articles were combined into a single file and the percentages of the different SIEs for each Step were calculated. The most frequently used tense/modality for each Step is given in bold. Infrequently used types are not shown.)

	Presenting Step							
	Def	His	Phe	Tech	CR	Ela	Lit	LitH
Total No.	68	252	1122	232	196	105	588	52
Past		**67.06**	4.54	3.88	**40.31**	13.33	**51.19**	**90.38**
Pres	**91.18**	6.75	**85.66**	**69.40**	15.31	**72.38**	20.41	7.69
PresP		18.25	2.85	18.10	33.16	4.76	25.85	
PastP							0.68	
M	10.29	3.57	6.95	8.62	4.08	9.52	1.53	
PresC								

Table 26. Verb tense/modality in Presenting Steps (%).
(For further details, see Table 25)

4.7.1 Present tense

The present tense was found to be the most common of all Discourse Organizer Steps, with the percentages being especially high for the Aim and DiscourseSignal Steps.

	Evaluating Steps			Synthesizing Steps		
	HisH	AsFL	Gap	Hypo	Dev	Op
Total No.	15	110	231	576	12	509
Past	**46.67**	24.55	3.03	1.74		4.32
Pres	6.67	**50.91**	**70.13**	25.35		**60.51**
PresP	6.67	20.00	17.32	1.74		7.86
M	40.00	4.55	9.09	**71.18**		27.31

Table 27. Verb tense/modality in Evaluating and Synthesizing Steps (%).
(For further details, see Table 25)

Here are some examples of present tense usage in the Aim Step.

SRAM 7:1 — *My purpose in this essay* is to review the evidence for somatosensory plasticity in the adult mammalian brain.

IBLE 1:9 — *Here I discuss some of these studies,* with their various caveats, that provide new information aiding in the further biochemical and molecular dissection of the complex signaling systems regulating MAP kinases and RSK.

TGRA 4:2	*One of the purposes of this paper* is to present the broader data base that now exists.	

In the Move Presenting Steps, very high percentages of the present tense were found for the Define, Phenomenon and Elaboration Steps. Examples from these Steps are as follows:

SVAU 10:1	*The term "programmed cell death" is commonly used synonymously with apoptosis.*	Def
HNIC 4:3	*Also forming part of the aglycon is a trisulfide, which serves as the trigger.*	Phe
TGRA 18:1	*A second type of ethological isolation is a product of flower constancy such as occurs in bees, lepidopterans, and to a lesser extent in long-tongued flies.*	Ela

4.7.2 Past tense

The other major verb tense for the Presenting Step was the Past tense, as can be seen for the History and LiteratureHist Step examples given below.

HOLS 6:8	In 1987, the first global human genetic map, based on "restriction-fragment-length polymorphisms," *was published (8).*	His
SLAI 10:2	Thus two HNF-3 family members, XFKH1 and pintallavis, *were obtained* from cDNA libraries of early frog gastrulae and *were shown* to be expressed in [...]	LitH

The past tense is also relatively frequently used for the Literature and Current Research Step, as shown in the following examples:

Literature Steps

IEPS 11:2	Jones and Handreck (17), in their classic review, *distinguished* three groups of crop plants on the basis of the Si content of their tissues.
SVAL 10:3	cDNA clones encoding the 70kDa intermediate chain (49) *were shown* to map to the ODA6 locus and were used to deduce the primary sequence of its polypeptide product (47).

HKNU 26:9 However, one family with high penetrance for RCC *carried* a constitutional translocation with one breakpoint at 3pl4-21 far from the VHL gene (74).

Current Research Steps

HNIC 24:3 Designed enediynes *demonstrated* abilities to cleave DNA and *exhibited* selective cytotoxicity against tumor cells versus normal cells (39, 42).

IBEU 20:2 Industrial-scale production of human placental enzyme modified so as to expose more terminal mannose residues (Ceredase; alglucerase) *made* available sufficient material for prolonged clinical administration of large amounts of acid beta-glucosidase.

SRAM 21:5 Furthermore, the distribution of these points *was* not random (10).

The SRAM article was unusual as it reported a substantial amount of research being done by the author. This is reflected in the large percentage (28.48 %) of CurrentResearch Steps found for this article. One interesting note on the Past tense is that it was not used in the SVAU article, which describes what is currently known about physiological cell death. The past tense was also an infrequent choice in the IAME (4 instances), ITEM (2), SGOL (5), TAND (3), TBEN (6) and TPRI (5) texts.

4.7.3 Modal verb

The term 'modal verb' is used here to refer to not only modal auxiliary but also verbs which have a modal-like meaning, such as 'appear' and 'seem'. The modal verb is very important for the Hypothesize Step.

HKNU 21:8 The possibility *must* be considered that not all of the tumors mentioned involve one and the same gene.

SVAU 21:3 Eventually a malignant clone *could* be produced.

IBLE 0:4 Thus, these signaling proteins *may* participate in the regulation of a variety of cellular processes.

TBRE 19:7 Thus, it *appears* that unregulated expression of LDL receptors can affect diet response.

The modal verb was the main verb form for the Aim Step. Many of the instances were of *will* and *shall*. Interestingly, all of the following examples appeared at the end of a paragraph, as can be seen by the *x, xx,* or *xxx* notation.

HOLS 10:1 x	For the purposes of this review, progress in the Human Genome Project *will* be examined relative to these three broad aims.
ITEM 2:7 Rx	I *shall* also consider in this paper the hypotheses (i) that misincorporation promotes strand transfers (13) and (ii) that misincorporation accompanies strand transfers (14).
SLAI 3:4 xx	In this review we *will* summarize the recent biochemical, structural, and biologic information about this protein family.
SRAM 20:5 xxx	In this essay, I *will* briefly summarize our findings for these two patients and will also describe some preliminary results from a third patient (FA).

4.8 Discourse function of the grammatical subject (GS) and Step type

The most GS prominent feature was the high frequency of RWE (Real World Entity) as the grammatical subject in the Define and Phenomenon Steps, which is understandable because these Steps are descriptions of items in the material world.

Here are some examples of RWE as the GS.

HOLS 3:6	*These hormones* are named after the cell type that secretes them.	Def
IBEU 4:1	*Gaucher disease* is one of a group of lysosomal storage disorders.	Def
TSCH 23:2	*An N-terminal fragment* that extends to Lys-300 contains the nucleotide binding fold and is fully active for adenylate synthesis.	Phe

	Discourse Steps			
Step	Topic	Aim	Disc	Sum
Total No.	240	28	130	56
Dmacro		21.43		
Dmicro			10.77	
PDisP		**35.71**	21.54	
Dint	1.67			
RWE	**52.50**		**35.38**	**35.71**
RWP	16.67			17.86
RWM	6.25			
HOEmp	4.17			

Table 28. Discourse functions of the GS (grammatical subject) in Discourse Steps (%). (All Steps from all articles were combined into a single file and the percentages of the different Grammatical Subject discourse functions for each Step were calculated. The most frequently used function for each is indicated in bold. Infrequently used items are not shown)

Most of the other steps also use RWE as the GS. Notable exceptions are the PDisP (Discourse participant) for the Aim Step, PIntP (Interactive participant) for the LitH (LiteratureHist) Step, and HOEmp (Empty hypothesized/objectivized theme) for the Op (Opinion) Step. PDisP was observed in relatively high percentages for the Aim (35.71 %) and DiscourseOrg (21.54 %) Steps.

Here are some examples:

SLAI 3:4	In this review *we* will summarize the recent biochemical, structural, and biologic information about this protein family.	Aim
TBEN 2:4	*We* will briefly discuss these results and propose a model for [...]	Aim
ITEM 4:4	*I* shall discuss, in relation to the strand-transfer hypothesis, [...]	Disc
IKEY 1:4	*I* refer to the contemporary disciplines, rather than to those of the 19th century.	Disc

	Presenting Steps							
	Def	His	Phe	Tech	CR	Ela	Lit	LitH
Total No.	68	252	1122	232	196	105	588	52
PIntP		7.54					15.31	**25.00**
PDisP					16.33			
Dint							5.44	
OIC							1.53	
RWE	**76.47**	**39.68**	**82.19**	**46.12**	**48.47**	**54.29**	**50.00**	
RWP		18.25	8.99	30.60	9.69	11.43	13.10	
RWM		6.75	1.16	3.02		6.67	1.87	
RWT							0.85	
Demp					4.08		1.36	
HOEmp							1.87	
RWat		5.16					1.19	
HOHv							1.36	
Oth							1.53	

Table 29. Discourse functions of the GS (grammatical subject) in Presenting Steps (%). (For further explanation, see Table 26)

	Evaluating Steps			Synthesizing Steps		
Step	HisH	AsFL	Gap	Hypo	Dev	Op
Total No.	15	110	231	576	12	509
PDisP			6.93	3.82		
Dint		8.18				3.73
OIC		13.64		3.30		4.52
RWE		**20.00**	**32.47**	**51.91**		
RWP		8.18	10.82	13.02		
RWM			6.49	2.26		
Demp				1.91		2.95
HOEmp			13.42	8.16		**10.41**
DDE				1.04		
HOHv				2.95		8.64
Oth				4.69		3.54

Table 30. Discourse functions of the GS (grammatical subject) in Evaluating and Synthesizing Steps (%). (For further explanation, see Table 26)

The PintP function was most frequently observed for the LiteratureHist Step (25.00 %) and the Literature Step (15.31 %). Most of the examples for this GS for the LiteratureHist Step were from HBRO:

HBRO 16:5	Based on these observations, *Kosterlitz and his colleagues* (30) proposed that a fourth type of opioid receptor, the delta receptor, must be present in the vas deferens.
HBRO 6:1	In 1806, *Serturner (8, 9)* isolated the active ingredient in opium and named it morphine after the god of dreams, Morpheus.
HNIC	*Masamune and coworkers (3) and Wong and Sondheimer (4)* also observed the cycloaromatization of 10-membered ring enediynes.

The HOEmp function was found in 13.42 % of the Gap Steps and 10.41 % of the Opinion Steps, being the most frequent type for the latter step.

Gap Step examples

SVAU 10:6	To add to the confusion, *there* are numerous reports of actinomycin D and cycloheximide themselves triggering apoptosis (S4-60).
TGRA 46:4	*It* may be difficult to say which mode is operating in any given case.
IBEU 9:8	*It* is difficult to understand how this state of affairs came about, [...].
HOLS 0:10	*There* is as yet little indication that the infusion of technology from outside biology into the Human Genome Project has been effectively stimulated.
SGOL 17:6	*There* also is no evidence for the existence of a PSID-like protein.

Opinion Step examples

TGRA 29:2	*It* has seemed best to me to group them in the four modes presented above, but alternative groupings into three or five types are also possible.
SLEN 18:2	*It* will be of great interest to see the effect of this disruption on the growth characteristics of mice or at least murine embryos.
HKNU 25:3	Still, *there* has been considerable progress in recent years, with a promise of more to come in the near future.
IBEU 18:2	Yet, *it* can be extraordinarily helpful in improving the quality of life of the patient and thus is not to be disdained.

The Dmacro (Macro discourse entity) function was another important one for the Aim Step, accounting for 21.43 % of the Grammatical Subjects. Here are some examples:

HNIC 0:4 *This review* summarizes the main contributions to the field, with particular emphasis on work from our laboratories.

TPRI 0:3 *This review* provides a perspective on these two types of neurodegenerative disorders.

TGRA 6:3 Thus *this paper* will examine the diverse origins of floral isolation.

4.9 Steps and review types

From Tables 31 to 35, it is clear which articles used the highest percentages of which Step type. The averages for the different review categories are given in italics. The tables show the Step distribution for each article as well as which articles tended to use which Steps.

As can be seen from these tables, the Phenomenon Step was the most frequently used in all review types. The Synthesizing Moves were more frequently found in the Status Quo, Theory and Issue reviews than in the History Reviews. The frequent use of the Hypothesize Step by the Theory reviews and the Opinion Step by the Issue reviews could be expected from the expressed aims of these articles.

	Discourse Organizing Steps								
	Topic	Aim	Disc	Sum		Topic	Aim	Disc	Sum
HBRO	3.51	0.00	0.00	1.75	TAND	0.00	0.79	3.17	0.00
HKNU	9.00	0.00	1.00	0.67	TBEN	1.48	0.74	2.96	0.00
HNIC	3.29	0.66	7.24	0.66	TBRE	4.79	0.00	0.00	6.38
HOLS	2.97	0.42	0.00	0.42	TGRA	7.79	0.32	6.82	0.32
					TPRI	9.23	1.54	0.00	1.54
					TSCH	2.90	0.72	2.17	2.17
Ave	*4.69*	*0.27*	*2.06*	*0.88*		*4.37*	*0.69*	*2.52*	*1.74*

	Discourse Organizing Steps								
	Topic	Aim	Disc	Sum		Topic	Aim	Disc	Sum
SCLE	7.66	0.85	1.28	0.43	IAME	10.61	0.00	0.51	0.00
SGOL	1.00	1.00	3.00	0.00	IBEU	6.76	0.97	0.97	0.97
SLAI	1.20	1.20	4.82	1.20	IBLE	2.94	0.98	1.96	2.94
SLEN	0.00	0.00	0.88	0.88	IEPS	11.27	0.36	4.73	5.09
SSHU	1.89	1.42	0.47	0.47	IKEY	3.29	0.47	13.15	1.41
SVAL	11.90	0.79	0.00	0.79	ITEM	6.84	0.85	4.27	0.85
SVAU	8.78	0.00	0.00	1.35					
SRAM	1.32	0.66	5.63	0.00					
SWIL	6.93	0.43	1.30	1.73					
Ave	*4.52*	*0.71*	*1.93*	*0.76*		*6.95*	*0.61*	*4.26*	*1.88*

Table 31. Discourse Organizing Steps and review types (%).
(When a unit was assigned more than one Step, only the first one only was counted for the calculations for this table)

	Presenting Steps							
	Def	His	Phe	Tech	CR	Ela	Lit	LitH
HBRO	1.75	31.58	27.19	0.00	0.00	0.00	5.26	20.18
HKNU	0.67	3.67	40.00	1.00	0.00	0.33	11.67	0.33
HNIC	0.00	28.95	13.82	1.32	3.95	0.66	0.00	13.82
HOLS	2.12	16.10	8.05	25.42	7.20	0.42	6.78	0.00
Ave	*1.13*	*20.07*	*22.26*	*6.93*	*2.79*	*0.35*	*5.93*	*8.58*
SCLE	0.85	10.64	9.79	15.32	5.11	0.43	8.51	0.00
SGOL	3.00	0.50	30.50	1.00	3.50	2.00	6.00	0.00
SLAI	0.00	8.43	36.14	1.20	3.61	0.00	13.25	3.61
SLEN	6.14	1.75	44.74	0.00	1.75	3.51	13.16	0.00
SSHU	0.47	6.13	3.77	34.43	4.72	0.94	26.89	0.00
SVAL	2.38	0.00	25.40	0.00	0.79	0.00	21.43	0.00
SVAU	2.70	0.68	39.19	0.00	2.03	1.35	0.68	0.00
SRAM	0.00	3.31	1.66	4.64	28.48	1.66	17.22	0.00
SWIL	2.16	3.90	20.78	5.19	6.49	2.60	13.42	0.00
Ave	*1.97*	*3.93*	*23.55*	*6.87*	*6.28*	*1.39*	*13.32*	*0.40*

Table 32. Presenting Steps for History and Status Quo reviews (%).

	Presenting Steps							
Step	Def	His	Phe	Tech	CR	Ela	Lit	LitH
TAND	4.76	1.59	26.98	0.00	3.97	0.00	10.32	0.00
TBEN	0.00	2.96	33.33	1.48	3.70	0.00	8.89	0.00
TBRE	4.26	0.00	19.15	8.51	4.26	0.00	37.23	0.00
TGRA	0.65	1.62	45.78	0.32	0.00	3.25	5.19	0.00
TPRI	6.15	0.00	32.31	0.00	0.00	1.54	15.38	1.54
TSCH	4.35	0.72	43.48	0.00	0.00	10.87	7.97	0.00
Ave	*3.36*	*1.15*	*33.51*	*1.72*	*1.99*	*2.61*	*14.17*	*0.26*
IAME	0.00	1.01	46.97	0.51	0.51	3.54	5.56	0.00
IBEU	0.48	11.11	32.85	0.97	4.35	0.00	9.66	0.00
IBLE	0.00	1.96	8.82	0.98	0.98	0.00	22.55	2.94
IEPS	0.36	2.55	15.27	1.45	0.36	4.00	28.00	0.00
IKEY	0.47	4.23	9.39	0.00	0.00	12.68	15.96	0.00
ITEM	0.85	0.00	41.03	6.84	2.56	1.71	5.13	0.00
Ave	*0.36*	*3.48*	*25.72*	*1.79*	*1.46*	*3.65*	*14.48*	*0.49*

Table 33. Presenting Steps for Theory and Issue reviews (%).

The Literature (Lit) Step or the LiteratureHist Step was found in all reviews, with the Lit Step being often used in most Status Quo Reviews. The exception was the SVAU article (0.68 %). It is not that the SVAU review did not use the Lit Step but rather that literature references were often included in other Steps, as can be seen from the following examples:

SVAU 1:4	*Descriptive and morphological studies have demonstrated the widespread occurrence and importance of cell deaths that occur as part of normal physiology (for reviews, see refs. 1-3).*	AsFL
SVAU 8:1	*The term "apoptosis" is often used to describe the physiological death of mammalian cells (1, 2).*	Def
SVAU 20:2	*For example, there are many examples of growth-factor deprivation causing apoptosis (73-75, 88-90)*	Ela/Lit
SVAU 21:5	*That this can, indeed, occur has been confirmed in transgenic mice that constitutively express bcl-2 in their lymphocytes (71, 72, 102, 103).*	Hypo

	Evaluating			Synthesizing		
Step	HisH	AsFL	Gap	Hypo	Dev	Op
HBRO	3.51	0.00	0.88	1.75	0.00	2.63
HKNU	0.00	1.33	5.67	15.33	0.00	9.33
HNIC	5.26	5.26	0.00	3.95	0.00	11.18
HOLS	0.42	0.42	10.59	2.97	0.00	15.68
Ave	*2.30*	*1.76*	*4.28*	*6.00*	*0.00*	*9.71*
SCLE	0.00	6.81	10.64	4.26	0.00	17.45
SGOL	0.00	5.00	6.50	23.50	0.00	13.50
SLAI	0.00	2.41	1.20	12.05	0.00	9.64
SLEN	0.00	0.00	7.02	14.91	0.00	5.26
SSHU	0.00	0.94	5.66	7.55	0.00	4.25
SVAL	0.00	0.79	3.97	12.70	0.00	19.05
SVAU	0.00	4.05	4.05	33.78	0.00	2.03
SRAM	0.00	5.96	4.97	14.57	0.00	9.93
SWIL	0.00	1.30	10.39	7.79	4.33	11.26
Ave	*0.00*	*3.03*	*6.04*	*14.57*	*0.48*	*10.26*

Table 34. Evaluating and Synthesizing Steps for History and Status Quo reviews (%).

	Evaluating			Synthesizing		
Step	HisH	AsFL	Gap	Hypo	Dev	Op
TAND	0.00	3.17	2.38	***39.68***	1.59	1.59
TBEN	0.00	3.70	7.41	25.19	0.00	8.15
TBRE	0.00	4.79	2.13	6.38	0.00	2.13
TGRA	0.00	0.32	6.17	10.39	0.00	11.04
TPRI	0.00	0.00	4.62	12.31	0.00	13.85
TSCH	0.00	0.00	1.45	18.12	0.00	5.07
Ave	*0.00*	*2.00*	*4.02*	*18.68*	*0.26*	*6.97*
IAME	0.00	3.03	2.02	21.72	0.00	4.04
IBEU	0.00	1.93	6.76	6.28	0.00	15.94
IBLE	1.96	5.88	8.82	***27.45***	0.00	8.82
IEPS	0.00	1.82	2.18	7.64	0.00	14.91
IKEY	0.00	0.00	2.35	0.94	0.00	***35.68***
ITEM	0.00	0.00	0.00	15.38	0.00	13.68
Ave	*0.33*	*2.11*	*3.69*	*13.23*	*0.00*	*15.51*

Table 35. Evaluating and Synthesizing Steps for History and Status Quo reviews (%).

Evidence for the relative freedom of expression in review articles was evident from exceptions to the general trends. For example, among the Status quo reviews, SRAM is unique in that it used a large number of CurrentResearch Steps. This review was unusual because it introduced a substantial amount of new information from studies currently in progress. As mentioned by most of the expert informants, review articles do not usually introduce new information. However, in a rapidly evolving field, a writer, who is usually a leading researcher, may decide that this is necessary. Another example is the SWIL article, which had 24 personal communication references. A possible explanation for this was provided by the informant (Newbury) who said that the writer may have wanted to get the latest data 'in print' as soon as possible. The fact that this review was published indicated that this was condoned by discourse community and, in fact, another informant (Wood) even praised this review as one that would be good for offering writing target guidelines.

4.9.1 History reviews

The History reviews showed a high percentage of History Steps, which could be expected. The Literature Step percentages were rather low but this was because there were many reference statements classified as LiteratureHist rather than as Literature.

HBRO 10:2	*Goldstein et al. (14) suggested that radiolabeled drugs might be used to demonstrate the existence of these receptors and to characterize them.*
HNIC 20:6	*Nicolaou et al. (65) were thus able to report the first total synthesis of the oligosaccharide portion of calicheamicin gamma1I toward the end of 1990.*

For the HKNU review, the Phenomenon Step (40.00 %) was the most frequently used, probably because this review catalogues various antioncogenes, or tumor suppressor genes, related to human cancers. The aim of the article is expressed as follows:

HKNU 4:4 *These discoveries have illuminated our understanding of human cancer considerably and have revealed a group of genes that is important in cellular and developmental biology, making this a propitious time to review progress on the subject and to take note of some future prospects for extending this understanding.*

This HKNU article also makes rather frequent use of the Hypothesize (Hypo) Step (15.33 %) because the author's hypotheses are expressed after the description of the various antioncogenes. Here are some examples of Hypo Steps:

HKNU 3:5 *If the 50 or so different hereditary cancers all arise by this mechanism, then there should be a corresponding number of antioncogenes.*

HKNU 6:2 *Since the number of transformable cells derived from the normal target cells, the embryonic retinoblasts, is of the order of magnitude of 107, the probability that a second, tumor-forming event will occur can be estimated as no less than 3 x 10-7 per cell division (5).*

HKNU 8:3 *Perhaps its action is circumvented by other regulatory mechanisms in tissues other than retina.*

For the HOLS review, the Technology step (25.42 %) was the most frequently used because this review describes advances in techniques that made possible the Human Genome Project:

HOLS 0:1 *The Human Genome Project in the United States is now well underway.*

HOLS 0:2 *Its programmatic direction was largely set by a National Research Council report issued in 1988.*

HOLS 0:3 *The broad framework supplied by this report has survived almost unchanged despite an upheaval in the technology of genome analysis.*

4.9.2 Status quo reviews

Seven of the ten Status quo reviews showed the highest percentage for the Phenomenon Step. The exceptions were SCLE which had instances of the 17.45 % Opinion Step, SSHU which had 34.43 % of the Technology Step, and SRAM, which as discussed above, had 28.48 % of the CurrentResearch Step. In the case of SCLE, the author states in the abstract that there are both problems and opportunities related to "Chloroplast gene sequences and the study of plant evolution" (title), which probably led to the high usage of the Opinion Step, similar to the Issue reviews. However, this article was classified as a Status quo review based on the statement of aim in the last sentence of the first paragraph.

SCLE 0:3	*The full analysis of this large data set poses many problems and opportunities for plant evolutionary biologists and for biostatisticians.*
SCLE 1:7	*The purpose of this review is to give a brief account of the history and status of this effort.*

The SSHU article had a large number of Tech Steps because it was about the "Nuclear magnetic resonance imaging and spectroscopy of human brain function". After describing the development of the various technologies, the review tries to describe their applications.

SSHU 0:1	*The techniques of in vivo magnetic resonance (MR) imaging and spectroscopy have been established over the past two decades.*
SSHU 0:2	*Recent applications of these methods to study human brain function have become a rapidly growing area of research.*

The second most frequent step for the Status quo reviews tended to be the Literature Step, except for SVAU, which has been discussed above, and the Hypothesize (Hypo) Step. SVAU also uses many Hypo Steps (33.78%), as can be expected from its title, "Toward an understanding of the molecular mechanisms of physiological cell death", which suggests that the author will be exploring the body of knowledge available. This article was classified as a Status quo review, partly based on the headings used to mark off sections of the text body; as suggested that there would be descriptions of known information which could be used to achieve the aim of the article.

SVAU 6:0	Dysregulation of Cell Death and Disease.
SVAU 6:0	Morphological Analysis of Physiological Cell Death.
SVAU 12:0	Genetic Analysis of Physiological Cell Death.

Another review using a relatively high percentage of Hypo Steps (23.50%) was SGOL. The aim is clearly stated in the abstract:

SGOL 0:6	*This review summarizes the current state of knowledge in comparative reaction center biochemistry between prokaryotic bacteria, cyanobacteria, and green plants.*

In the text itself, the author does more than merely summarize, as can be seen from some of the following examples:

SGOL 5:7	*This high degree of homology strongly suggests that the sequences arose by duplication and divergence from a single ancestral gene.*
SGOL 7:2	*This implies that the RC in C. limicola is probably a homodimer instead of a heterodimer.*
SGOL 7:4	*The strong implication is that the heliobacterial and green sulfur bacterial RCs and PSI shared a common ancestor.*

4.9.3 Theory reviews

The Theory reviews had the highest average percentages of the Phe (Phenomenon) Step (33.51 %) and the Def (Define) Step (3.36 %) in comparison with the other types of reviews. This is probably because phenomena need to be specified and definitions need to be explicit when trying to present a theory or model. Below are some examples of the Def Step.

TAND 4:1	*Each caveola is a dynamic piece of membrane that is either open for receiving and releasing material or closed for processing, storage, and delivery to the cell (11).*
TBRE 1:2	*There are five types of lipoprotein particles: from smallest to largest, HDL, LDL, IDL, VLDL (high, low, intermediate, and very low density lipoproteins), and chylomicrons.*
TPRI 1:4	*PHFs are composed principally of abnormally phosphorylated isoforms of tau (11-13), a low molecular weight microtubule-associated protein.*
TPRI 0:1	*#Alzheimer disease and the prion disorders/spongiform encephalopathies share many common features.*
TSCH 1:1	*#The genetic code is an algorithm that relates each amino acid to one or more trinucleotide sequences.*

Another characteristic of Theory reviews was a relatively high percentage of Hypo (Hypothesize) Steps (average 18.68 %). This is understandable as the Aim Steps of these reviews indicate that known findings will be discussed and then synthesized into a new model, for example:

TGRA 6:3	*Thus this paper will examine the diverse origins of floral isolation.*
TPRI 0:3	*This review provides a perspective on these two types of neurodegenerative disorders.*
TAND 3:1	*In this article, I outline the data that support a role for caveolae in cell signaling.*
TBEN 0:4	*We discuss the results that support this possibility and propose a model for synaptic vesicle docking and fusion [...]*

4.9.4 Issue reviews

The Issue reviews use a relatively high percentage of Opinion Steps in comparison with the other types of reviews. This can be expected from what is expressed in the Aim Steps of these reviews, which note that 'implications', 'questions' and 'caveats' will be discussed:

IBEU 7:3	It is these newer developments and their *implications* that are the major focus of this brief review.
IBLE 1:9	Here I discuss some of these studies, with their various *caveats*, that provide new information aiding in the further biochemical and molecular dissection of the complex signaling systems regulating MAP kinases and RSK.
ITEM 2:7	I shall also consider in this paper the *hypotheses* (i) that misincorporation promotes strand transfers (13) and (ii) that misincorporation accompanies strand transfers (14).

This chapter began with a presentation of the various Moves and Steps used in the different sections of the different types of reviews. However, the Move boundaries were difficult to identify and in order to devise useful writing support tools for nonnative English speakers, more detailed analyses were needed to identify microlevel features. The basic unit chosen for analysis was the orthographic sentence (clauses and phrases were used when necessary), with each sentence being assigned a Step function. To clarify the features of these Steps, in this chapter, we have examined SIE (sentence initial element), discourse functions of the GS (grammatical subject), and Vb (verb tense/modality). Finally, the Steps themselves and their relative frequency of use were used to identify the features of the different types of reviews. Next, let us turn our focus to the features and the key words and phrases of the Discourse Organizer and Move Steps in preparation for the development of writing support tools.

5. Discourse organizer and move steps of review articles

> You might be chuckling when reading this because my message represents a typical Japanese flow of thought.
>
> [e-mail message from a Japanese colleague, February 27, 1998]

5.1 Identifying the features

What did my colleague mean in the above e-mail message? Some of the answers may lie in the use/non-use of Steps, their ordering and the use of collocations not usually used by native speakers in the discourse community. In Chapter 4, the Steps were analyzed into their components which included SIE (sentence initial element), GS (grammatical subject), and Vb (verb tense/modality), and the relationships between the different Steps and review type were examined. In this chapter, the Steps will be examined in various ways with help from *Wordsmith*, a bank of linguistic analysis tools, including a concordance and a word list comparison function, by Mike Scott (version 2.0, Oxford University Press). Presented below are KeyWord lists, words used with high frequency, and examples, together with other information which could be useful for writing support measures which will be discussed in Chapter 7.

The KeyWord list function is described in the Help Menu of the *Wordsmith* program as follows:

> This is a program for identifying the 'key' words in one or more texts. Key words are those whose frequency is unusually high in comparison with some norm [...]. The program compares two pre-existing word-lists, which must have been created using the WordList tool. One of these is assumed to be a large word-list which will act as a reference file. The other is the word-list based on one text which you want to study. The aim is to find out which words

characterise the text you're most interested in, which is automatically assumed to be the smaller of the two texts chosen. The larger will provide background data for reference comparison.

The KeyWords lists that follow were obtained by comparing word lists prepared for each Step, including all instances of that Step in one text file together with their identifying tag and the word list for the entire corpus of reviews which were in individual text files but were combined for the analysis. Here is a portion of the text file for the Def (Define) Step:

SWIL 38:4	Ribozymes can be visualized as "warheaded" antisense RNAs; however, the length and base composition of the two arms will affect the hybridization on/off rates at a given temperature and, hence, the kinetics and efficiency of RNA cleavage.
TGRA 2:2	Mechanical isolation can occur when two or more plant species have different flower structures that reduce or prevent interspecific pollination.
SGOL 0:2	These can be classified as "quinone type" (type 11) photosystems because the terminal electron acceptor is a mobile quinone molecule.
SVAU 10:1	The term "programmed cell death" is commonly used synonymously with apoptosis.

As described in Chapter 4, these Steps were first tentatively proposed then presented to the specialist informants. Based on their feedback and the features described in Chapter 4, these Steps were collected into text files for the analyses discussed in this Chapter which are aimed at trying to elucidate the Step features to create useful writing support tools for nonnative speakers of English.

Other identification markers were the *R* and *x, xx, xxx* used, respectively, to indicate the presence of a reference citation and the end of a paragraph or section.

5.2 Discourse Organizer Steps

Part of the list from Chapter 4, Section 4.5 is reproduced here.

Discourse Organizer Steps: Organizing

Topic	Introducing a topic
Aim	Expressing aim
Disc	Organizing discourse – referring to discourse elements, using rhetorical questions, imperative verb
Sum	Summarizing

5.2.1 Discourse Organizer Step: Topic – Introducing a topic

This Discourse Organizer Step almost always appears as a first sentence of a paragraph which is very likely to be the first paragraph of a section. This is to be expected as it is used to introduce a new topic. The Keywords search did not reveal any significant key words other than *plant* which, being a content word, was considered to have arisen due to the high incidence of bioscience papers in the corpus uses. This absence of key words could be expected because the Topic would vary with the article. The Plot function showed a generally even distribution throughout a text when it was used.

Here are some examples of sentences of this Step. Given in bold type are words and phrases, not directly related to the content, that are considered to signal the Step type. In some cases, e.g. HNIC 1:1, the sentence as a whole was considered to serve as a signal.

IAME 28:1	***The epidemiological evidence and the guidelines*** *of the National Cancer Institute and the National Research Council/National Academy of Sciences* ***suggest that*** *at least two fruits and three vegetables per day is a desirable intake.*
SRAM 33:1	*#The upper-limb amputees as well as the digit amputees spontaneously volunteered* ***the following intriguing observation.***
HNIC 1:1	***Nature continually serves to provide scientists with new ideas, often provoking us to reexamine previous studies and leading us in new directions.***

TBEN 17:1 ***Many of the fundamental** biochemical **processes essential for** living organisms are conserved across most species.*

The Topic Step is also very often combined with a Move Step which presents field-specific information.

TPRI 6:1	*#Several in vitro systems **have been used to explore** the influences, toxic or trophic, of Abeta on neurons (67).*	Topic/Lit
SVAU 18:1	***The final step in the process of** programmed cell death in C. elegans is dissolution of the nucleus of the corpse.*	Topic/Phe
IKEY 0:1	*#Policy on population and environment in the United States and abroad **has been vacillating,** unsure of its course.*	Topic/Gap
HBRO 10:1	***By the mid-1960s, it was becoming clear that** the actions of opiate agonists, antagonists, and mixed agonist-antagonists **could best be explained by** actions on multiple opiate receptors.*	Topic/His

5.2.2 Discourse Organizer Step: Aim – Expressing aim

The expression of aim appears very early in the paper, with 8/27 instances in the abstract and 16/27 instances within the first seven paragraphs or before the main body of the review text. Table 36 presents the KeyWords search results for this Aim step. Words such as *review*, *this*, and *discuss* can be expected. The *xxx* indicates that this step often appeared at the end of a section. The table is followed by concordance lines for these key words.

WORD	FREQ.	AIMTXT.TXT %	FREQ.	PNASX.TXT	KEYNESS %	P
REVIEW	10	1.33	36	0.03	55.1	0
THIS	24	3.19	615	0.48	49.8	0
XXX	7	0.93	51	0.04	29.9	0
DISCUSS	4	0.53	6	27.7	0	

Table 36. KeyWords for Aim Step.

Here are some sample concordance lines for 'review' in the Aim Step:

0 1 x For the purposes of this **review**, progress in the Human
sfers (14). SLAI 3 4 xx In this **review** we will summarize the rec
It is the purpose of this article to **review** recent developments in the
are the major focus of this brief **review**. TPRI 0 3 xxx This review
brief review. TPRI 0 3 xxx This **review** provides a perspective on
ative disorders. HNIC 0 4 This **review** summarizes the main
laboratories. SWIL 0 5 This **review** attempts to consider, in a
osed rules. SGOL 0 6 xx This **review** summarizes the current s
My purpose in this essay is to **review** the evidence for somatose
CLE 1 7 x The purpose of this **review** is to give a brief account

Sample concordance lines for 'this' for the Aim Step are as follows:

ations that are the major focus of **this** brief review. TPRI 0 3
f this brief review. TPRI 0 3 xxx **This** review provides a
nerative disorders. HNIC 0 4 **This** review summarizes the
OLS 10 1 x For the purposes of **this** review, progress in the
broad aims. TGRA 6 3 xx Thus **this** paper will examine the
A 4 2 One of the purposes of **this** paper is to present the
SHU 4 1 It is the purpose of **this** article to show how recen
rbcL sequences. TAND 3 1 In **this** article, I outline the data
ns. SRAM 7 1 My purpose in **this** essay is to review the
tein family. SRAM 20 5 xxx In **this** essay, I will briefly
green plants. SVAL 3 11 Rxx **This** work is reviewed here

Of the 24 instances, 22 (91.7%) refer to the aim of the paper as a whole, with 8 using the word *review*, 4 *paper*, 3 *article*, 2 *essay*, 1 *work*. Some refer to the overall aim while not referring directly to the text itself.

Examples of concordance lines for *discuss* are:

f the medium. TSCH 6 1 We also **discuss** what is known
FA). TBEN 2 4 xxx We will briefly **discuss** these results
ed to be answered. IBLE 1 9 x Here I **discuss** some of these
in cell signaling. TBEN 0 4 xxx We **discuss** the results that

The *xxx* symbol was also listed among the key words showing that the Aim Step often appears at the end of a paragraph or section. The fact that this symbol appeared in the KeyWords list indicates that the Aim Step is more significantly found at the end of a section than other steps. This clear statement of aim coming at what can be considered the end of the introduction section finds parallel in Swales' Establishing a Niche step (Swales 1990a) which is found at the end of the introduction section in a research article. Swales was referring to scientists describing the need for their specific research as a contribution to advancing field knowledge. On the other hand, in a review article, the authors seem to be describing the rationale for their evaluation and synthesis of this knowledge to promote further advance of not only their own field-specific community but a broader scientific community. These are some examples of the sentence starts at the end of a paragraph or section:

in structure and function. IBEU 7 3	**xxx**	It is these newer
and status of this effort. IBEU 3 2	**xxx**	The study of Gaucher
out this protein family. SRAM 20 5	**xxx**	In this essay, I will
n of these substrates. SSHU 4 2	**xxx**	We sketch, at this early
m a third patient (FA). TBEN 2 4	**xxx**	We will briefly discuss
olae in cell signaling. TBEN 0 4	**xxx**	We discuss the results
cus of this brief review. TPRI 0 3	**xxx**	This review provides a

Here are some complete sentence examples of the Aim Step which refer to the paper as a whole:

HOLS 10:1	x	***For the purposes of this review, progress in the Human Genome Project will be examined relative to these three broad aims.***
ITEM 2:7	x	***I shall also consider in this paper the hypotheses** (i) that misincorporation promotes strand transfers (13) and (ii) that misincorporation accompanies strand transfers (14).*
SRAM 20:5	xxx	***In this essay, I will briefly summarize our findings for these** two patients **and will also describe some preliminary results from** a third patient (FA).*
TGRA 6:3	xx	***Thus this paper will examine*** the diverse origins of floral isolation.

This Aim Step is also characterized by the use of the present tense verb (21/27 instances, 77.78%) or *will* or *shall* used in the future sense and the use of the Dmacro/Dmicro discourse function of the grammatical subject.

IEPS 3:4 ***As is customary, concentrations are given on a*** *dry weight* ***basis and expressed as*** *percentages for the macronutrient and ppm (mg/kg) for the micronutrient elements.*

HNIC 0:5 ***Highlights include studies of*** *the Bergman reaction,* ***which is central to the mechanism of*** *action of enediynes, the design and chemical synthesis of a number of these systems, and biological studies with selected molecules.*

TGRA 17:7 ***For ease of reference we will call this*** *common mode the Aquilegia type of ethological isolation.*

ITEM 4:4 ***I shall discuss, in relation to*** *the strand-transfer hypothesis, minus-strand and plus-strand DNA primer transfers and each of these types of genetic variation.*

Examples of Dmacro/Dmicro as the subject are:

IEPS 21:2 ***For this reason the present review is largely confined to*** *work with nutrient solutions.*

HNIC 12:9 ***Fig. 7 exhibits*** *a number of other designed enediynes with triggering devices and modulated reactivity synthesized in these laboratories [51 (44), 52 (unpublished results), 53 (45), 54 (45), 55 (43), and 56-58 (46)].*

IBEU 11:5 ***Table 3 shows*** *the haplotype of patients with the three mutations that are most common in the Jewish population.*

5.2.3 Discourse Organizer Step: Disc (DiscourseSignal) – Organizing discourse

This Step is characterized by the key words presented in Table 37. The IKEY text had a significantly large number of instances of this step probably because it was originally presented as a speech.

N	WORD	FREQ.	DISCTXT.TXT %	FREQ.	PNASX.TXT %	KEYNESS	P
1	IKEY	28	0.92	238	0.19	42.7	0
2	WHAT	13	0.43	47	0.04	37.2	0
3	TABLE	12	0.39	39	0.03	36.4	0
4	HERE	9	0.3	24	0.02	30.1	0

Table 37. KeyWords for Disc Step.

The significant frequency of *what* for the DiscourseSignal Step seems to point to the author trying to directly involve the audience in the dialogue. Of the 13 instances found, five were from the SRAM text which, as has been mentioned above, is unusual because of the ample number of references to current research in the author's laboratory. Here are some sample concordance lines:

ese processes.) SCLE 11 1 First, **what** is a neutral
s follows: IKEY 49 2 That is not **what** I have tried to do.
With regard to our empirical results, **what** is novel may be
pecies are disappearing anyhow, so **what** matter if humans
rct with each other? SRAM 46 6 **What** would happen if
ancy of bumblebees. SGOL 10 1 **What** was the nature
the beta-elimination? TGRA 47 3 **What** are the
ts of the hybridization? IKEY 9 1 # **What** has all this to do
touch and hearing? SRAM 37 1 # **What** is the actual

The frequent use of the word *here* in the Disc Step also shows the author's awareness of the audience, as can be seen from these examples:

llular communication. SCLE 18 2 xx **Here**, we outline the basic
nt. TGRA 51 3 x We are interested **here** in a possible parallel
RA 18 6 a The result of interest to us **here** is that different
ple beyond the 4 billion or so already **here**. IKEY 13 4 x Label
tRNA acceptor stems are described **here**. TSCH 7 4
arize briefly these behavior patterns **here**. ITEM 22 3 Ux [It
5 4 x Such an analysis is presented **here**. TGRA 64 1
prevent description of these results **here**. IEPS 3 2 The
ral are the findings we have reported **here**? SRAM 49 1

The concordance lines for *table* is of interest from the collocation viewpoint as this word is frequently used. It shows the following verbs used to refer to this text entity: *show, list, refer, depict,* and *base.*

erred to throughout the text and in **Table** 1. IBLE 11 4
e discussed below. IBEU 11 5 **Table** 3 shows the
its title (37). SLAI 2 2 Shown in **Table** 1 are representative
s reviewed in ref. 23. SLAI 9 4 **Table** 1 lists the known
64)]. SGOL 2 1 As depicted in **Table** 1, the identities of
he cases known to me are listed in **Table** 1. TGRA 9 1 One
ed that all but a few of the cases in **Table** 1 occur, as would be
er that heading.) IEPS 3 6 The **table** is based on analyses
hese results here. IEPS 3 2 The **table** lists all the essential

This DiscourseSignal Step is characterized by the predominant use of the present tense verb (95/135, 70.37%), as can be expected because the author(s) is trying to guide the reader through the text. A unique feature of this Step rarely found in the other Steps is the use of imperative sentences (11/130, 8.46%) and questions (30/130, 23.07%), as in the following examples:

IKEY 46:4	***Suppose that*** *the doubling of population had some indirect good effects that could offset the harm of doubling.*
TAND 16:5	***Imagine*** *two closely apposed cells where patches of closed caveolae on one cell are adjacent to clusters of purinoceptors on the other (Fig. 3, Unstimulated).*
TGRA 48:1	***Let us restrict this discussion to*** *floral isolation composed of the Salvia type of mechanical isolation combined with the Aquilegia type of ethological isolation.*
SRAM 46:6	***What would happen if*** *an upper-limb amputee is asked to move his phantom fingers?*
HNIC 15:5	***Is there*** *a prevailing biological mechanism in certain cell types which facilitates the beta-elimination?*

There were also instances of moodless clauses which were in response to a question or comment posed in the previous sentence (note the pairs below in which the second is not a complete sentence). This was found in IKEY which was originally presented as a speech.

IKEY 9:1	*What has all this to do with the differences in attitudes toward growth?*
IKEY 9:2	*A great deal.*

The DiscourseSignal Step was most frequently used in IKEY, TGRA, SRAM, IEPS and HNIC, based on the Frequency list data. The x mark (multiple x marks indicate section endings) in the table indicates the end of a paragraph, which serves as an ending comment.

TGRA 29:4	x	***It may be useful to summarize briefly these behavior patterns here.***
IEPS 21:4	x	***In this section on*** *growth,* ***however, some experiments with*** *soil-grown plants* ***will be included because they dramatically illus-***

trate the importance of *Si for the growth of* ***at least*** *some species* ***under field conditions.***

SRAM 48:1		*How general are the findings we have reported here?*
HNIC 17:2	Rxx	***A selection of these model compounds covering the rather wide spectrum of*** *structural types synthesized in this field is shown in Fig. 9 [63 (SS), 64 (56), 65 (57), 66 (58), 67 (59), and 68 (60)].*

However, the DiscourseSignal Step appeared even more often at the beginning of a paragraph, 33/130 (25.8 %) in sentences such as the following:

IKEY 23:1	*#Species are disappearing anyhow, so what matter if humans cause a few more to vanish?*
TAND 15:1	***Thus far in the discussion,*** *caveolae have been portrayed as storage and processing sites that ultimately deliver messages to the inside of the cell.*
TBEN 10:1	***Our current model for*** *synaptic vesicle docking and fusion* ***is presented in Fig. 2.***

The Frequency list showed that *we* (14) and *I* (13) were often used. The usage of *we* in TGRA and IKEY was not to refer to the author himself (TGRA, Verne Grant; IKEY, Nathan Keyfitz) but to both the writer and the audience. In the case of *I*, however, the pronoun specifically referred to the author. These pronouns were used in the following ways:

TGRA 51:3	x	***We are interested here in*** *a possible parallel process involving floral isolating barriers.*
IKEY 12:3		*Do we want growth enough to risk irreversible damage to the ecosystem that is our ultimate support?*
SRAM 7:3	xxxx	*I* ***will focus, instead, on our own*** *behavioral* ***work*** *and on the MEG correlates of somatosensory reorganization in humans.*
IKEY 25:8	N	***I cite*** *Lee and Devore only to give the flavor of the long-range preoccupation of a discipline – anthropology – that straddles biological and social science.*

Here are examples of the DiscourseSignal Step used together with other Steps.

IEPS 12:2		***For a discussion of*** *mechanistic* ***aspects of*** *its transport,* ***which are beyond the scope of this paper, the review by*** *Raven (14)* ***is unexcelled.***	Disc/ AsFL
HKNU 9:11	x	***As we shall see later, the apparent explanation is that*** *for these other tumors, other controls on growth must be overcome; RBI is not the only gene that must mutate.*	Disc/ Hypo
SCLE 27:1		***As indicated in the preceding discussion, the analysis*** *of the complete rbcL data base* ***is just beginning.***	Disc/ Op
SRAM		***#With regard to our empirical results,*** *what is novel may be summarized as follows:*	Disc/ Sum

5.2.4 Discourse Organizer Step: Sum – Summarizing

The KeyWords list for the Sum step yielded *thus* and *I*. The *I* was part of a name A-I which appeared in the TBRE text. Therefore the only significant word was *thus*. Eleven of the 12 instances of *thus* appeared at the beginning of the sentence and eight were in sentences which end the paragraph. Here are some examples:

HKNU 18:8	xx	***Thus, there is specificity not only for*** *an organ* ***but also for the degree of differentiation*** *within that organ.*
HBRO 20:9	x	***Thus, unlike*** *receptor candidates cloned earlier, it behaves just as one might have expected it to.*
TBRE 14:9	xx	***Thus,*** *half-normal E expression in the mouse* ***is nearly sufficient for*** *normal lipoprotein metabolism.*

The one instance in which *thus* is not sentence-initial was the following:

IEPS 4:5	xx	*Si* ***thus figures as a major mineral constituent of*** *plants.*

Overall, this Sum Step uses the present tense most often, 32/56 (57.14%) and is frequently found at the end of a paragraph, 26/56 (46.43%).

HKNU 17:11	x	*TP53* ***plays an important role in*** *these latter two tissue categories, but the second category is the principal target for oncogenesis in LFS.*

IEPS 2:5	x	*Si* ***therefore looms large as one of the major constituents of*** *the soil solution in contact with plant roots.*
IEPS 14:6	x	***There are fairly distinct differences in*** *the distribution and characteristics of silica depositions between C3 and C4 plants (69, 77).*

Interestingly, *conclude* and its related words appeared in only three instances.

TPRI 11:1		***In conclusion,*** *AD and the prion disorders* ***show many parallels.***
TSCH 22:1		***Similar conclusions have been obtained with*** *the class I methionine tRNA synthetase.*
IEPS 10:1	xx	***In concluding this section let it be noted that*** *Si is an essential element for higher animals (45).*

As for the word *summary* and its related words, only one instance was found for this Sum Step.

SSHU 25:1	***In summary,*** *MRS* ***methods now exist for measuring*** *anaerobic glycolysis, for determining kinetics and rates of glucose transport across the blood-brain barrier, and* ***for deriving*** *from the glutamate turnover the tricarboxylic acid cycle flux.*

A *Wordsmith* search showed that *summary* and its related words are used in other Steps, especially the Aim Step as follows:

SRAM 36:1		***#With regard to our empirical results,*** *what is novel may be summarized as follows:*	Disc/ Sum
SRAM 20:5	xxx	***In this essay, I will briefly summarize our findings*** *for these two patients and* ***will also describe some preliminary results from*** *a third patient (FA).*	Aim
SLAI 3:4	xx	***In this review we will summarize the recent biochemical, structural, and biologic information about*** *this protein family.*	Aim
SGOL 0:6	xx	***This review summarizes the current state of knowledge in*** *comparative reaction center biochemistry between prokaryotic bacteria, cyanobacteria, and green plants.*	Aim
HNIC 0:4		***This review summarizes the main contributions to the field, with particular emphasis on work from our laboratories.***	Aim

The Sum Step is often found together with other moves, 29/56 (51.79%), seven of which express the author's opinion. Here are examples of the Sum Step used with the Opinion Step. The *x* (and multiples of *x*) indicate the last sentence of a paragraph or section.

SWIL 0:7	xxxx	*Any unregulated superimposition of interfering protein or nucleic acid species* ***can, therefore, result in*** *an apparently virus-resistant plant phenotype.*	Sum/ Op
IEPS 43:1		***Together, this evidence amounts to a powerful case for the importance of*** *Si in plant biology.*	Sum/ Op
IBEU	x	***As a consequence, there is relatively little relationship between*** *the severity of disease and whether or not a patient receives treatment.*	Sum/ Op
IEPS 2:5	x	*Si* ***therefore looms large as one of the major constituents of*** *the soil solution in contact with plant roots.*	Sum/ Op
IBLE 11:1		***As indicated by the title of this review, there is great risk in trying to describe*** *the ever-expanding, complex signaling network involved in the regulation of MAP kinases.*	Sum/ Op
IEPS 15:4	xx	***This is an aspect of*** *biomineralization and* ***should be viewed in that context*** *(53, 79-81).*	Sum/ Op
SVAL 23:1		***At present it is difficult to make complete sense out of*** *the disparate but fascinating properties of the components of the Glued or dynactin complex.*	Sum/ Op/ Gap

Although there are eight instances of *it*, only three are used as HOEmp (Empty H and O theme):

IEPS 10:1	xx	***In concluding this section let it be noted that*** *Si is an essential element for higher animals (45).*
SVAL 23:1		***At present it is difficult to make complete sense out of*** *the disparate but fascinating properties of the components of the Glued or dynactin complex.*
SCLE 24:12	xxx	***It is clear, however, that*** *most analyses understate the sources and magnitude of error and* ***hence*** *overstate the precision of phylogenetic estimates.*

5.3 Move Steps – Presenting

The move steps related to Presenting are given here once more:

Def	Giving a definition
His	Describing historical events
Phe	Describing generally accepted phenomena
Tech	Describing key technology, method, tool
CR	Describing current research – not specific, general and current
Ela	Presenting examples, explanation
Lit	Introducing related literature
LitH	Introducing historical literature

5.3.1 Move Step, presenting: Def – Giving a definition

The Define Step most often uses the copular in the present tense with the SIE (sentence initial element) being the grammatical subject with a real world entity function. It frequently was the first sentence in its paragraph and was found most often in the Theory and Status quo reviews, TBRE (8 instances), SLEN (7), and SGOL, TAND, and TSCH (6).

The KeyWord search for the Define Step yielded two words, *is* and *electron,* the latter being a content word appearing in several texts (SWIL, SGOL, SLEN). The Plot function showed that this Step tends to be scattered in the text for the History Reviews, is hardly used in the Issue Reviews, and tends to appear in the first half of Status quo and Theory reviews.

The high frequency of the copula *is* (37/68, 54.41%) could be expected. The plural *are* (8/68, 11.76%) was also often found.

TSCH 1:1	*The genetic code* ***is an algorithm*** *that relates each amino acid to one or more trinucleotide sequences.*
HKNU 7:3	*Loss of heterozygosity of at least some syntenic markers* ***is a feature of all of these mechanisms except the first.***
SVAL 1:2	*These proteins are microtubule-activated ATPases, which convert chemical energy into mechanical energy.*

ITEM 3:1	*Retroviruses are a family of animal viruses that alternate their genetic material between RNA in the virion and DNA in the infected cell (15, 16).*

This Define step appeared as the first sentence in the paragraph in 25 of the 68 instances found (36.76%).

SLEN 0:1	*The interferons are **a family of** secreted, multifunctional proteins which are components of the defenses of vertebrates against viral, bacterial, and parasitic infections and certain tumors.*
TPRI 0:1	*Alzheimer disease and the prion disorders/spongiform encephalopathies **share many common features**.*
HOLS 2:1	*The human genome is the genetic material in human egg and sperm cells (i. e., germ cells), which contain 3 x 109 base pairs (bp) of DNA.*
SGOL 4:1	*PSI **is considered** an "iron-sulfur-type" RC in which the bound primary quinone, Al, donates its electron to an iron-sulfur cluster (probably Fx) **instead of** to the secondary quinone (reviewed in ref. 9).*
SVAU 8:1	***The term** "apoptosis" **is often used to describe** the physiological death of mammalian cells (1, 2).*

5.3.2 Move Step, presenting: His – Describing historical events

The History Step was most frequently found, as could be expected, in the History reviews, which accounted for 51.19% (129 instances), with 26.98% (68) for the Status quo reviews, 17.06% (43) for the Issue ones, and 4.76% (12) for the Theory ones.

This History Step is characterized by the frequent use of the past tense, 170/252 (67.46%), which differs from the verb usage of the other Steps. It is also frequently used with a reference citation, as shown below. Here are examples of History Step sentences:

HKNU 30:1	R	*The DCC gene **was identified following LOH studies that showed** frequent losses for markers on chromosome 18q (87).*
IAME 19:3	R	***Earlier papers have called attention to a number of previously neglected physiological antioxidants**, including urate, bilirubin, carnosine, and ubiquinol (111-114).*

SCLE 12:1	R	***Early work in molecular evolution seemed to support** a linear relationship between clock time and the accumulation of mutational change (e.g., see ref. 22).*
TAND 7:1	R	*Insulin **was the first** hormone suspected of using inositol phosphoglycan (IPG) or a molecule derived from IPG as a second messenger (19-21).*

N	WORD	FREQ.	HISTXT.TXT %	FREQ.	PNASX.TXT %	KEYNESS	P
1	WAS	86	1.2	377	0.3	101.1	0
2	HNIC	43	0.6	156	0.12	61.4	0
3	HBRO	36	0.5	114	0.09	58.1	0
4	FIRST	28	0.39	114	0.09	35.6	0
5	WERE	40	0.56	231	0.18	32.9	0
6	AGLY-CON	11	0.15	15	0.01	30.6	0
7	HOLS	38	0.53	237	0.19	27.8	0
8	SYN-THESIS	18	0.25	63	0.05	26.5	0
9	ARE	11	0.15	818	0.64	37.6	0
10	IS	25	0.35	1,491	1.17	56	0

Table 38. KeyWords for the History Step.

One interesting finding for the History Step was the frequent use of the word *first*, with 11 instances of use with verbs and 17 with nouns. Here are concordance lines of the collocations with a verb:

the sequence of the rbcL gene was	**first** accomplished by
uence, reasonably efficient methods	**first** appeared in 1977
the aglycon is not as daunting as at	**first** appeared. HKNU
hese molecules thus constituted the	**first** designed
an more than a decade ago when I	**first** encountered
1 Plasmalemmal caveolae were	**first** identified as an
In fact, it was Claude Bernard who	**first** investigated the
The idea of a molecular clock was	**first** proposed by
ble activity of these antibiotics were	**first** studied (2). HNIC
as 1901(4) IBEU 5 4 R Although	**first** thought to be
States, the Human Genome Project	**first** took clear form in

Here are usage examples of *first* with a noun:

tosis. TAND 7 1 R Insulin was the	**first** hormone
rebral blood flow. SSHU 3 2 R The	**first** in vivo NMR
HNIC 5 1 R Dynemicin A (7), the	**first** member of the
inol (111-114). HOLS 14 3 aR The	**first** paper on the PCR
ymes was the development of the	**first** practical method of
on University (St. Louis) led to the	**first** report of
and esperamicin Al (6, Fig. 1), the	**first** representatives of
ed dynorphin. HBRO 12 5 x The	**first** such species
the full functionality to achieve the	**first** synthesis of
licheamicin gammal1 (5) (71), the	**first** total synthesis of

5.3.3 Move Step, presenting: Phe – Describing generally accepted phenomena

The Phe Step was the most frequently found type, totaling 1,123 instances. It was found with significantly high frequency in TGRA, IAME, HOLS, IKEY, SCLE, SSHU and SRAM.

The verb tense used was mainly the present tense, accounting for 85.57%, with the next closest being modal verbs at a mere 6.95%. Another high ranking item in the frequency table was the R symbol indicating the presence of a citation in the sentence. It is interesting that references are cited even when the present tense is used in what seems to be a simple statement of fact.

Here are some examples of the use of the present tense in the Phe Step.

HBRO 13:2	*In mammals there* **are** *three such precursors-proenkephalin (24), prodynorphin (25), and proopiomelanocortin (26).*
IAME 7:4	*Smoking* **is** *a risk factor for heart disease as well as a wide variety of cancers in addition to lung cancer (35-38).*
SGOL 11:4	*A [4Fe-4S] cluster* **forms** *in both proteins, but it is easily converted to a [3Fe-4S] cluster under oxidizing conditions (28, 29).*
TGRA 20:6	*Lapses of flower constancy in which an individual bumblebee crosses over from one Pedicularis species to the other* **occur** *but are rare (9, 32).*

N	WORD	FREQ.	PHETXT.TXT %	FREQ.	PNASX.TXT %	KEY-NESS	P
1	TGRA	141	0.48	319	0.25	38.4	0
2	ARE	275	0.94	818	0.64	28.4	0
3	CELLS	112	0.38	270	0.21	25.5	0
4	SPECIES	90	0.31	203	0.16	24.7	0.000001
5	IAME	93	0.32	213	0.17	24.5	0.000001
6	HOLS	19	0.07	237	0.19	26.3	0
7	IKEY	19	0.07	238	0.19	26.6	0
8	SI	13	0.04	197	0.15	27.5	0
9	SCLE	20	0.07	251	0.2	28.1	0
10	WE	7	0.02	155	0.12	29.9	0
11	WERE	16	0.05	231	0.18	30.6	0
12	MAY	20	0.07	271	0.21	33.4	0
13	BEEN	40	0.14	432	0.34	38.8	0
14	THAT	187	0.64	1,349	1.06	47.6	0
15	SSHU	8	0.03	219	0.17	48.1	0
16	BE	68	0.23	677	0.53	53	0
17	WAS	24	0.08	377	0.3	54.7	0
18	SRAM	6	0.02	327	0.26	95.2	0

Table 39. KeyWords for the Phe Step.

The Frequency list showed many instances of *is* and *are* and also of *not*. Here are some examples of combinations of these words.

IEPS 37:1 *Rice is **not only** grown on heavily leached mineral soils low in available Si **but** on largely organic low-Si soils **as well**.*

HKNU 22:4 ***Such is not the case for** MEN2, whose mutations predispose heterozygotes to two neural crest-derived tumors, pheochromocytoma and medullary carcinoma of the thyroid.*

SSHU 27:2 *In functional MRI using Hb-based contrast, the activation-related signal changes depend upon rCBF, rCBV, and rCMRO2 in a way **that is not understood at present quantitatively**.*

TBEN 10:4 ***However**, the process of synaptic vesicle fusion **is not** constitutive **but rather** is tightly regulated by calcium.*

Another feature of the Phe Step was the predominant use of the subject in the SIE position (730 instances, 65.00%). The next most frequently used SIE was the CondR with 95 instances (8.46%). Of the Phe with the subject in the SIE position, 62.51% were RWE.

TGRA 24:3	*The plant group may be a tribe, genus, species group, or species.*
SWIL 22:3	*Plant lines containing all other gene constructs were susceptible to PVX inoculation.*
IAME 6:7	*Chronic infections are* ***particularly prevalent*** *in third-world countries (see below).*

One interesting KeyWord finding for this step was the use of *we* which appeared with a significantly higher frequency than in the total corpus, although there were only seven instances. All except the last concordance line are from Issue review texts, which may suggest a stronger author presence when trying to draw attention to the existence of some issue.

ed. IKEY 24 4 xx For economists,	**we** should hesitate to do
by joint replacement. IBEU 29 9 a	**We** can identify a fetus
e grossest form of interference, and	**we** should act to bring it
excreted in the urine. IAME 3 4	**We** estimate that the
se mutations in 100 Jewish patients	**we** have studied (36).
nts. IAME 11 1 R Smoking, which	**we** and others argue is
; TGRA 12 2 In the first, which	**we** will call the Salvia

5.3.4 Move Step, presenting: Tech – Describing key technology, method, tool

Words such as *method(s)* and *image* appear with high frequency in this step, as can be expected. Also, SSHU, HOLS and SCLE used more of this step than the other texts. The four reviews which had the largest number of Tech Steps were the following and their titles:

SSHU	Nuclear magnetic resonance imaging and spectroscopy of human brain function.
HOLS	The human genome project.
SCLE	Chloroplast gene sequences and the study of plant evolution.
TBRE	Transgenic mouse models of lipoprotein metabolism and atherosclerosis.

As suggested by their titles, technology and methods play key roles in these reviews even though they are from different categories. In the case of HOLS, the development of a project was closely related to technological advances in the field.

Frequent use of the present tense verb in 159/230 (69.13%) is also a characteristic of this Step. This is understandable as this move is used to describe technology important to the field. Here are some sample concordance lines for *method*:

S 13 2 R PCR, which is essentially a	**method** of in vitro cloning,
ntent mapping provides a convenient	**method** of establishing
S in vivo builds upon well-established	**methods**, which show
dividual subjects. SSHU 2 2 Both	**methods** take advantage
man brain responses to stimuli by this	**method** has already
steady fixation.) SSHU 0 6 The	**methods** have high
ivity dependent. SSHU 0 5 These	**methods** have allowed
stematics. SCLE 23 5 The bootstrap	**method** has been widely
cess in physical mapping, PCR-based	**methods** have also
l imaging. SSHU 7 5 In the FLASH	**method**, each line is
. l). SSHU 7 4 R Multiple excitation	**methods** require at least
ational process, SCLE 23 4 MLE	**methods** provide a

N	WORD	FREQ.	TECHTXT. TXT %	FREQ.	PNASX. TXT %	KEY-NESS	P
1	SSHU	74	1.16	219	0.17	141.1	0
2	HOLS	60	0.94	237	0.19	89.7	0
3	METHODS	23	0.36	72	0.06	41.9	0
4	METHOD	16	0.25	30	0.02	40.9	0
5	MAGNETIC	15	0.24	26	0.02	40	0
6	MRI	17	0.27	43	0.03	36.1	0
7	SCLE	38	0.6	251	0.2	30.9	0
8	SIGNAL	17	0.27	60	0.05	28	0
9	TREES	11	0.17	21	0.02	27.8	0
10	RF	9	0.14	12	27.3	0	
11	EXCITATION	8	0.13	9	26.1	0	
12	DNA	38	0.6	288	0.23	24.7	0.000001
13	IMAGE	10	0.16	20	0.02	24.6	0.000001
14	PP	9	0.14	15	0.01	24.5	0.000001

Table 40. KeyWords for the Tech Step.

Some refer to nonspecific methods, while others (79.5 %) use a demonstrative *(this)* or a specific determiner *(both)* or specifically name the method, showing that references to the type of method tend to be precisely stated.

A word closely related to *method, technique(s)* was used in sentences such as the following:

TBRE 5:2	***In addition, recently, the technique of*** *homologous recombination in embryonic stem cells* ***has been used to*** *knock out lipoprotein transport genes.*
HOLS 23:1	***While*** *the PCR,* ***together with such new techniques as*** *YAC cloning, FISH, and radiation-hybrid mapping,* ***has led to a surge of success in*** *physical mapping, PCR-based methods have also transformed genetic mapping*
IBEU 2:4	***At the same time,*** *the techniques of molecular biology have made new treatment options available.*
SRAM 46:1	***Another promising technique*** *(50)* ***for studying*** *reorganization of maps in humans involves the use of transcranial magnetic stimulation.*

There also were numerous references to citations, as found in the Phe Step, even when the description seemed to be of an accepted fact or condition.

IEPS 43:12	R	***In this laboratory,*** *1 mM Na2SiO3 is included in a modified Hoagland solution (12).*
HOLS 13:2	R	*PCR,* ***which is essentially a method of*** *in vitro cloning, allows the amplification of specific DNA molecules in vitro through cycles of enzymatic DNA synthesis (ref. 4, pp. 7985).*
SSHU 7:4	R	*Multiple excitation methods require at least one RF pulse for each line of the image matrix (12, 14).*

Other instances of references were frequently made to specific developments, either historical or contemporary.

HOLS 18:2	R	***One such development has been the introduction of*** *the yeast artificial-chromosome (YAC) cloning system,* ***first described in*** *1987 (ref. 4, pp. 590-592; ref. 24).*
IAME 3:3	R	***Methods have been developed to assay*** *several of these excised damaged bases in the urine of rodents and humans (1, 16),* ***almost all of which appear as*** *the free base from repair by glycosylases.*

SRAM 43:4	R	***Recent advances in** MEG **have been made possible with the advent of** large array magnetometers (>30 detectors) and the understanding of the physics and mathematics of the measurements of cortical electrical activity (45-48).*

5.3.5 Move Step, presenting: CR – Describing current research

The KeyWords search for the CurrentResearch step yielded the list shown in Tablc 41. As is clear from the table, many of the terms were from SRAM which included a substantial amount of explanation of research currently in progress. Three other reviews also had a number of CurrentResearch Steps:

SRAM	*Behavioral and magnetoencephalographic correlates of plasticity in the adult human brain.*
HOLS	*The human genome project.*
SCLE	*Chloroplast gene sequences and the study of plant evolution.*
SWIL	*Strategies to protect crop plants against viruses: Pathogen-derived resistance blossoms.*

The SRAM text was interesting as a review because it carried a large amount of information from ongoing research in the author's laboratory, accounting for 86/197 of the CurrentResearch Steps, or 46.65%. Therefore, many of the words in the high-frequency table come from the SRAM article. Such a high frequency of CR Steps was not found in the other reviews, which did however, carry references to personal communications (total, 35 instances; 3 for T texts, 30 for S texts, of which 24 were for SWIL, and 3 for I texts).

N	WORD	FREQ.	CRTXT. TXT %	FREQ.	PNASX. TXT %	KEY-NESS	P
1	SRAM	86	1.55	327	0.26	152.5	0
2	PHANTOM	32	0.58	59	0.05	90.4	0
3	HIS	23	0.41	44	0.03	63.8	0
4	FACE	23	0.41	46	0.04	62.3	0
5	POINTS	21	0.38	37	0.03	60.7	0
6	WE	35	0.63	155	0.12	54.2	0
7	AMPUTATION	17	0.31	27	0.02	51.6	0
8	HAND	20	0.36	55	0.04	44.8	0
9	ARM	18	0.32	48	0.04	41.1	0
10	HE	14	0.25	26	0.02	39.4	0
11	WATER	15	0.27	32	0.03	39.2	0
12	PATIENT	16	0.29	42	0.03	36.9	0
13	WARM	10	0.18	10	36.7	0	
14	FELT	10	0.18	13	0.01	33.2	0
15	SHE	10	0.18	13	0.01	33.2	0
16	COLD	10	0.18	13	0.01	33.2	0
17	SENSATION	9	0.16	10	31.8	0	
18	DIGIT	10	0.18	18	0.01	28.6	0
19	FINGER	11	0.2	25	0.02	27.7	0
20	SENSATIONS	12	0.22	37	0.03	24.9	0.000001
21	ON	51	0.92	520	0.41	24.8	0.000001
22	RECENT	15	0.27	62	0.05	24.7	0.000001
23	WAS	41	0.74	377	0.3	24.5	0.000001
24	TIP	7	0.13	8	24.4	0.000001	
25	IS	24	0.43	1,491	1.17	33.3	0

Table 41. KeyWords for the CurrentResearch Step.

As mentioned in Chapter 3, the SWIL article was considered by the expert informant to be unusual because of the abundance of personal communication references. The relatively greater number of personal communication references in S texts than in other types is understandable as the author(s) is trying to describe the current situation in the field and would probably wish to include as much recent information available as possible but might be hampered by the time lag in getting research published. In the case of the SWIL review, the informant said there was about a 9-month time lag. Here are some examples of personal communication references found with the CurrentResearch Step:

SWIL 4:4	P	***Several groups are attempting*** *CPMP against important virus diseases of rice (rice tungro spherical virus and rice stripe tenuivirus; J. W. Davies and P. Tien, personal communications) and maize.*
TBEN 8:8	P	***Genetic studies have revealed that*** *this open reading frame is one of two genes, 5501 and 5502, that are capable of acting as multicopy suppressors of a temperature-sensitive allele of SEC1 (S. Keranen, personal communication).*
SSHU 13:3	P	***Schneider (personal communication) has begun a detailed analysis of*** *the human visual cortex.*

Table 41 shows that *recent* was a key word with significantly high frequency. The concordance lines for *recent* for this step show that it is often used for the status quo and theory reviews and at the start of a sentence followed by words such as *studies* (6) *and evidence* (3). It also often appears with a reference citation *R*, and sometimes with *P* (personal communication) or *U* (unpublished).

m the angiosperms, SSHU 0 2 **Recent** applications of
s and strategies. SRAM 0 1 xxxxx **Recent** behavioral and
ing area of research. TAND 20 2 R **Recent** evidence
oducts, respectively. SWIL 14 2 RP **Recent** field tests with
e footing with PSI (10). SSHU 21 7 R **Recent** improvements in
DNA-binding structure. SLEN 1 2 R **Recent** reports have
(unpublished data). SGOL 6 1 R The **recent** sequencing of
tory (70). SLAI 9 6 However more **recent** studies (see
uced oxidative damage. TBEN 8 1 **Recent** studies of the
ogenies (55). SCLE 31 5 R A second **recent** study has
us challenge. SWIL 21 4 P However, **recent** work has

Another interesting feature of this CurrentResearch Step was the use of the present continuous tense which was rarely found in other Steps.

TAND 2:2	***In addition to being a site of entry for*** *small molecules and ions by potocytosis,* ***evidence is accumulating that they also play a major role in*** *coordinating the interaction of the cell with its environment.*

SCLE 27:2		***In the meantime,*** *sequences are being added to the data base* ***at an ever-increasing rate.***
HNIC 23:3	x	***Research in this field is continuing unabated*** *in chemistry, biology, and medicine, with some compounds already in clinical trials.*
SWIL 4:4	P	***Several groups are attempting*** *CPMP against important virus diseases of rice (rice tungro spherical virus and rice stripe tenuivirus; J. W. Davies and P. Tien, personal communications) and maize.*

5.3.6 Move Step, presenting: Ela – Presenting examples, explanation

The KeyWords search for the Elaborate Step revealed only two items, *example* and *IKEY,* of which *example* could be expected. The word *example* appears as *for example* in 15 of the 20 examples in the ELA corpus and 11 are sentence initial.

SVAU 20:2	***For example,*** *there are many examples of growth-factor deprivation causing apoptosis (73-75, 88-90).*
SSHU 31:3	***For example,*** *activation during encoding and retrieval phases of a memory task can be observed independently at different brain sites.*
IAME 28:5	*Folic acid,* ***for example,*** *is required for the synthesis of the nucleotides in DNA.*
TGRA 15:1	***Other examples of*** *the Salvia type of isolation are found in Aquilegia, Fuchsia, Impatiens, Ipomopsis, Penstemon, Monarda, and Lobelia (Table 1).*

The other key word was IKEY, which indicated that this step was used most frequently in this review which was originally a speech and therefore may have included more elaboration than the other texts (27 instances in comparison with the 15 of the next most frequent text, TSCH). A reference citation was given in 19 instances (19/105, 18.10%).

5.3.7 Move Step, presenting: Lit – Introducing related literature

The KeyWords search for the Literature Step yielded many terms. Items with *show* appeared in 21 instances with the present perfect verb (44.68 %), 12 with the past (25.53 %), 3 with the past perfect (6.38 %), and only once with the present tense (2.12%). The frequent use of the present perfect suggests the linking of previous findings and results to what is being discussed in the text.

Here are example concordance lines for *shown*:

The transfected PKR cDNA was also **shown** to decrease
A have been identified (110,111) and **shown** to reduce
a and delta receptors have also been **shown** to inhibit the
(10). SLAI 10 4 Activin has been **shown** to induce axis
E 28 6 R Inadequate intake has been **shown** to cause
SSHU 13 2 R Importantly, it has been **shown** that conventional
study of stimulation of the rat paw had **shown** an increase of
SHU 19 2 R Earlier animal studies had **shown** that lH MRS can
A-l levels and HDL size, as previously **shown**, and also showe
sequence comparison, it was quickly **shown** by gene-transfer
else. SVAL 1 5 R It was subsequently **shown** to be structurally
R The normal cDNA of RBI was then **shown** to cause
TBRE 4 2 R Dietary cholesterol was **shown** to increase
er plants already referred to (31) was **shown** to be due to the
r tobacco ringspot virus (TRSV) were **shown** to provide

Examining *recently* shows that it appears as a Sentence Initial Element (SIE) in seven instances and in five as *more/most recently*. It is also used in seven instances in the following construction: *has/have recently* V-*ed*. Here are some examples with *recently*.

ease of IPG from cells. TAND 9 4 R **Recently** a set of
ubgroup II (28). SWIL 38 6 aR Most **recently** in plant
n-inducible ribonuclease, was cloned **recently** (26). SLEN
may be more than one), LCFS2, has **recently** been mapped
68). IAME 25 2 RN Taylor (172) has **recently** reviewed the
esicle docking protein, syntaxin, has **recently** been identified
enosine monophosphate (AMP) has **recently** been found to
Block and her colleagues (93) have **recently** reviewed 172
of our filling in effect was observed **recently** by De Weerd et
bination in murine cells was reported **recently** (25). SLEN
somatotopic. SRAM 13 5 RNx More **recently**, Pettet and
arasitic RNAs." SWIL 41 5 Most **recently**, the CP of
ells (36, 37). HOLS 24 8 R Still more **recently**, a human

5.3.8 Move Step, presenting: LitH – Introducing historical literature

There were only 52 examples of LiteratureHist Step, with 23 examples for HBRO, one for HKNU, 21 for HNIC, 3 for IBLE, 3 for SLAI, and one for TPRI. As indicated by the KeyWords list, there is a significantly high frequency for HNIC. The high frequency of the Literature-Hist Step for history reviews is to be expected. The other three terms, *cleavage*, *opium*, and *opiate*, are topic-specific words.

N	WORD	FREQ.	LITTXT. TXT %	FREQ.	PNASX. TXT %	KEY-NESS	P
1	WAS	133	0.78	377	0.3	77.6	0
2	R	234	1.37	918	0.72	67.7	0
3	BEEN	133	0.78	432	0.34	60.1	0
4	ET	42	0.25	65	0.05	52.3	0
5	AL	43	0.25	72	0.06	49.7	0
6	SI	72	0.42	197	0.15	44.5	0
7	TBRE	70	0.41	197	0.15	41.3	0
8	MICE	45	0.26	100	0.08	37.7	0
9	WERE	75	0.44	231	0.18	37.6	0
10	SHOWN	37	0.22	76	0.06	34.2	0
11	STIMULATION	28	0.16	50	0.04	30.3	0
12	RN	34	0.2	75	0.06	28.7	0
13	RECENTLY	29	0.17	57	0.04	28.2	0
14	LEVELS	42	0.25	110	0.09	27.8	0
15	IEPS	77	0.45	284	0.22	26	0
16	TRANSGENIC	40	0.23	109	0.09	24.9	0.000001
17	INCREASE	31	0.18	72	0.06	24.5	0.000001
18	THE	795	4.65	7,067	5.55	24.8	0.000001
19	MAY	8	0.05	271	0.21	29.9	0
20	BE	41	0.24	677	0.53	31.3	0
21	DNA	8	0.05	288	0.23	33.2	0
22	ARE	47	0.28	818	0.64	41.7	0
23	IS	78	0.46	1,491	1.17	89.1	0

Table 42. KeyWords list for the Literature Step.
(R = reference citation; RN = reference citation includes author names)

N	WORD	FREQ.	LITHTXT. TXT %	FREQ.	PNASX. TXT %	KEYNESS	P
1	HBRO	23	1.51	114	0.09	83	0
2	HNIC	21	1.37	156	0.12	61.3	0
3	RN	10	0.65	75	0.06	29	0
4	CLEAVAGE	6	0.39	17	0.01	27.2	0
5	OPIUM	6	0.39	21	0.02	25.1	0.000001
6	OPIATE	5	0.33	11	24.7	0.000001	

Table 43. Key words for the LiteratureHist Step.
(*RN* = reference citation with author(s)' name(s))

5.4 Move Steps – Evaluating

Here are the Move Steps for Evaluating:

HisH Hypothesizing about historical events
AsFL Assessing findings, literature
Gap Describing a gap or limitation in the knowledge of the field

5.4.1 Move Step, evaluating: HisH – Hypothesizing about historical events

This HistoryHypo step had only two significantly high-frequency key words, *would* and *HNIC*, a history review. Here are examples of *would* concordance lines:

HNIC 19 5 bx and so there **would** be no opportunity
of molecular models indicated that it **would** be sterically
y the required beta-glycosidic linkage **would** be formed if it was
were so ambitious that they probably **would** have
l synthesis of calicheamicin gammal1 **would** not only require
nthesis would require a strategy which **would** provide large
HNIC 19 2 Such a daring endeavor **would** require careful
uction. HNIC 21 2 This synthesis **would** require a strategy

This step was mainly found in the History reviews but appeared in two instances in the IBLE review:

IBLE 1:3	***Several years ago*** an ~90-kDa cell cycle-regulated S6 protein kinase (***now referred to as*** pp90rsk or RSK) was identified and shown to be regulated by serine/threonine phosphorylation (***for review see*** refs. 1 and 2).
IBLE 1:4	***Soon after***, a mitogen-activated protein kinase [referred to as MAP kinase or extracellular signal-regulated kinase (ERK)] ***was identified that*** phosphorylated and partially reactivated dephosphorylated RSK (3).

5.4.2 Move Step, evaluating: AsFL – Assessing findings, literature

The authors present their view of other people's work with this move Step. The only word which appeared with significant frequency was *these*, which appeared in 12 (36.36%) instances in the sentence initial position to point to previously mentioned information. Words frequently used with *these* include *data, experiments, observations,* and *results*, as can be seen from the following examples:

ugh cell viability was not affected in	**these**	cases (49). HNIC
A. HNIC 16 2 RNx Notable among	**these**	contributions are
within the dicots. SCLE 29 1 b and	**these**	data have confirmed
In the subclass (29). SRAM 33 3 x	**These**	effects were
(13-16) and I will therefore mention	**these**	experiments only b
s 56-58 (Fig. 7) (46). HNIC 15 2	**These**	experiments
tor neurons (190, 191). IBEU 11 7	**These**	findings are
D 10 4 R Two recent reports now put	**these**	observations on a
lis) has further stimulated interest in	**these**	organisms as
t, and atherosclerosis. TBRE 22 4	**These**	results suggest
oncentrations of Ca2+ are present at	**these**	sites, which
TAND 13 9 x Taken together,	**these**	studies suggest that
ant, probably owing to the paucity of	**these**	substitution events
(12). SCLE 6 2 R It is precisely at	**these**	taxonomic levels
ch of these mutations. IEPS 6 6 R	**These**	values are on the

Here are examples of complete sentences using *these* and also *this* in the AssessFinding Step:

SCLE 28:4 ***In combination, these results suggest that*** *many of the major flowering plant lineages may have appeared in conjunction with a series of rapid evolutionary radiations early in angiosperm evolution.*

HNIC 16:2 ***Notable among these contributions are those of*** *Schreiber (59, Fig. 8) (51), Wender (60, Fig. 8) (52), Magnus (61, Fig. 8) (53), and Isobe (62, Fig. 8) (54).*

IBLE 8:8 ***These data are also consistent with the possibility that*** *MEKK and c-Raf can work independently or together to activate MEK by various agonists.*

TBEN 12:7 ***This result is consistent with the idea that*** *synaptotagmin is a negative regulator and* ***that it is not absolutely required for*** *membrane fusion.*

TBRE 20:3 ***Therefore, this model has not been useful in confirming the role of*** LRP in chylomicron-remnant metabolism in vivo.

IBLE 0:2 ***The focus of much of this information has centered around*** *the agonist-dependent activation of the mitogen-activated protein (MAP) kinases.*

With the AssessFinding Step, the names of the authors involved are usually not given unless the reference is a positive evaluation perhaps because of the possible confrontational stance, as in the following examples:

SRAM 45:5 *Although perfect symmetry between the two hemispheres in normal subjects is not guaranteed,* ***the results of*** *Yang et al. (48)* ***suggest that this is generally true.***

TAND 13:6 ***Sargiacomo et al. (56) now have evidence that*** *they are associated.*

HNIC 11:3 ***Particularly interesting were the observation of*** *Danishefsky and collaborators (37, 38) that such simple enediyne systems can simulate the DNA cleaving properties of calicheamicin gammal1 and the synthetic work of Magnus and coworkers (34, 35) which utilized acetylene cobalt complexes as intermediates to arrive at the targeted enediynes.*

HNIC ***Notable among these contributions are those of*** *Schreiber (59, Fig. 8) (51), Wender (60, Fig. 8) (52), Magnus (61, Fig. 8) (53), and Isobe (62, Fig. 8) (54).*

However, when there is a need to directly contradict another's work, the author(s) must be named. This is found in the following example from the issue-raising review, IEPS, "The anomaly of silicon in plant biology."

IEPS 7:10	***Thus the observation by*** *Miyake and Takahashi (25)* ***is an instance of*** *an element, Si in this case, correcting an unfavorable chemical condition in the medium,* ***and that, according to the classical definition for*** *essentiality* ***laid down by*** *Arnon and Stout (21), does not qualify the element for "essential" status.*

This AssessFinding Move Step was associated with other Steps in 20/110 instances (18.19%).

SRAM 59:4	xxxxx	The rapid occurrence of such precise, functionally effective sprouting ***has never before been documented in*** the adult human brain and it provides some grounds for optimism.	AsFL/ Dev
HNIC 11:2	RN	***Notable among the early contributions are the studies of*** Danishefsky (e.g., 35, Fig. 4) (31, 32), Kende (36, Fig. 4) (33), Magnus (e. g., 37, Fig. 4) (34, 35), and Schreiber (38, Fig. 4) (36) and their coworkers.	AsFL/ His
TAND 10:3	R	Histochemical ***methods have even indicated that*** high concentrations of Ca2+ are present at these sites, ***which suggests*** a storage role (36, 37).	AsFL/ Hypo
TAND 10:4	R	***Two recent reports now put these observations on a firrner footing***: a form of the inositol 1,4,5-trisphosphate-regulated Ca2+ channel (38) and a Ca2+-pump ATPase have been localized to caveolae (39).	AsFL/ Lit
SGOL 15:7	R	***A comparison of*** the g ***values*** and g-tensor orientations of the Chlorobium RC with PSI ***strengthens the argument that there is significant interaction between*** FA and F8 (4).	AsFL/ Op
HKNU 9:6		***Even more surprisingly***, several other tumors not typically observed in retinoblastoma survivors were mutated at RBI.	AsFL/ Phe
SCLE 28:4		***In combination, these results suggest that*** many of the major flowering plant lineages may have appeared in conjunction with a series of rapid evolutionary radiations early in angiosperm evolution.	AsFL/ Sum

5.4.3 Move Step, evaluating: Gap – Describing a gap or limitation in the knowledge of the field

This Gap Step is important in science in that it is used to point out where further research is needed. The Plot function showed clusters of this Step in HKNU and HOLS toward the end of the paper, SCLE after the middle, SWIL at the beginning, and toward the end of the paper in TGRA. Examination of the key words revealed the following useful collocations:

not	*not apparent, has/have not V-ed, it is not clear, we do not/it is not known, not well understood, not yet V-ed*
remain	*remains to be* V*-ed*
whether	*determin* whether, it is not known whether, question of whether*
no	*no evidence (for)*
difficult	*it is difficult to* V, *is/are difficult*

N	WORD	FREQ.	GAPTXT. TXT %	FREQ.	PNASX. TXT %	KEY-NESS	P
1	NOT	57	0.98	372	0.29	53.8	0
2	REMAINS	15	0.26	21	0.02	46.8	0
3	IS	126	2.16	1,491	1.17	37.1	0
4	WHETHER	16	0.27	41	0.03	36.1	0
5	NO	27	0.46	143	0.11	32.9	0
6	DIFFICULT	12	0.21	35	0.03	24.8	0.000001

Table 44. KeyWords list for Gap Step.

Here are some examples for the use of *not* in the Gap Step:

ad. HOLS 39 5 x However, it is **not** apparent that there has
However, this phenomenon is **not** applicable to all viruses
ociated with hormone levels are **not** as consistent and show
, and this switching is generally **not** available on clinical MRI
een MEK and MAP kinase may **not** be so simple. IBEU 29 11
a is rare and linkage analysis has **not** been conducted. HOLS
10 13 x Although this model has **not** been studied in detail, it
for lipoprotein metabolism have **not** been reported (16).
nance (ESR) characteristics have **not** been described. TBEN 9
mponents, such a connection has **not** been demonstrated.

to look for. HKNU 29 9 x It has **not** been established
Unfortunately, the Fx cluster has **not** been positively identified
patibility. TGRA 65 2 It is **not** clear how the Pedicual
4 a Unfortunately, the terms are **not** clearly defined, IBLE
TAND 13 5 These studies did **not** determine whether the
rther study. TGRA 33 2 We do **not** have enough evidence to
regulate their expression but are **not** included in the mRNA,
SVAU 19 4 b but at present it is **not** known whether the
IBLE 2 9 x Unfortunately, it is **not** so simple (Fig. 1 and
ed (60-62). SCLE 5 3 x This is **not** the case for most
SWIL 8 2 In most cases, we do **not** understand, and so
es are empirical, but the data are **not** wholly convincing.

Below are examples of *remain** in the Gap Step.

al identification of the gene still **remains** a formidable task
by which such maps are formed **remains** an interesting
his viral "cross-protection" have **remained** elusive and
r. HOLS 21 6 bx but the need **remains** for supplementary
11 xx Opportunities in this area **remain** large, posing
s level of structural redundancy **remains** obscure. SVAU 11 1
TGRA 5 1 A question which **remains** problematical is the
ol of eIF-2 phosphorylation still **remains** to be proven. SWIL 6
her causes. SRAM 36 9 xxx It **remains** to be seen, of course,
nctions in membrane trafficking **remains** to be established.
tigial or serve a regulatory role **remains** to be investigated.
n containing 2[4Fe-4S] clusters **remains** to be identified in
d Table 1). HKNU 23 4 b and it **remains** to be determined

Whether was another key word in the Gap Step.

It remains to be seen, of course, **whether** this latent
These studies did not determine **whether** the tyrosine
29 9 x It has not been established **whether** there is a
4 b but at present it is not known **whether** the activation of
in detail, it raises the question of **whether** other
find all mutations. IBLE 3 10 Rx **Whether** this 63-kDa
not have enough evidence to say **whether** the ornithophilous
U 37 6 b although it is uncertain **whether** they are changes
gical technologies. TBEN 3 8 **Whether** nonneuronal

Examples of *no* usage in the Gap Step included the following:

r than one-pass, partial sequences; **no** adequate solution
not been established. SCLE 24 11 **No** attempts have yet
less definitive (95, 96). IEPS 39 5 x **No** clear-cut correlation
the outset it is accurate to say that **no** consensus model
per heavy chain (45). SWIL 4 9 a **No** effort has been
n. SCLE 24 10 Currently, we have **no** estimates of the
rror; SGOL 171 6 There also is **no** evidence for the
HKNU 23 6 x Still, there is as yet **no** explanation for
data set. SVAL 7 2 ab There is still **no** indication as to why
t wholly convincing. IKEY 48 19 x **No** one knows what
s. HKNU 36 2 b but there has been **no** progress in
bal optimum. SCLE 24 8 a There are **no** reliable estimates
1 4 xxx To our knowledge, there are **no** such effects reported

Still another key word for the Gap Step was *difficult.*

s run to completion, it is usually **difficult** if not impossible to
lly possible but also theoretically **difficult**, improbable, and
c progenitor stem cells is a more **difficult** task, IBEU 35 4 R
ions on the natural blind spot are **difficult** to make because it
xpected. TGRA 46 4 It may be **difficult** to say which mode
peutic purposes. IBEU 9 8 It is **difficult** to understand how
ons. IBEU 14 2 What is more **difficult** to explain is the

The frequency list includes negative signaling words such as *not, no, but, however, although* and *unfortunately* display high frequencies. Here are some examples of how these words are used:

TGRA 33:2	***We do not have enough evidence to say whether*** *the ornithophilous and sphingophilous species of Diplacus are florally isolated* ***to any significant extent.***
SGOL 4:6	***The function of*** *the secondary quinone in PSI* ***is not known.***
HKNU 23:6	***Still, there is as yet no explanation for*** *pedigrees that show highly penetrant dominant inheritance of gliomas, even over three generations (for example, see ref. 68),* ***which may point to*** *initiating events in glioma formation and to a new antioncogene.*
SGOL 7:	***However, no*** *contemporary organism* ***has yet been discovered with*** *a quinone type homodimeric RC.*

SRAM 9:2	***Unfortunately, observations** on the natural blind spot **are difficult to make because** it is so far from the center of gaze, and patients with small, well circumscribed scotomas of cortical origin are not easy to come by.*

This move is also sometimes found in the second or later clause of the sentence, 32/230 (13.91 %).

IEPS 31:5	aR	***Effects of*** Si on the toxicity of other heavy metals such as Al ***are likely*** (97),	Hypo
IEPS 31:5	bx	***but information on the subject is scant.***	Gap
HOLS 21:6	a	YAC-based ***methods have led to greatly improved continuity***,	His
HOLS 21:6	bx	***but the need remains for*** supplementary methods to define the order and orientation of disconnected contigs along chromosomes.	Gap

5.5 Move Steps – Synthesizing

Move Steps for Synthesizing are the following:

Hypo	Hypothesizing – about specific phenomena: trying to explain
Dev	Explaining possibilities for further development and applications
Op	Giving opinion – expression of judgments of value, validity

5.5.1 Move Step, synthesizing: Hypo – Hypothesizing about specific phenomena

The Plot function showed a tendency for this Step to appear toward the ending of the paper in the Status quo reviews. Examination of the key words identified reveals useful collocations for the Hypo Step.

*suggest**	*These data suggest, limited/preliminary/a wealth of evidence suggests that, These findings suggest, These observations suggest that, results suggest that, strongly suggests, studies suggested, This suggests (that), we suggest*
likely	*is/are likely to* V*, most likely, It seems likely that*
appear	Real World Entity subject *appear to* V, *it appears that*
if	Very often used (78.72%) as the sentence initial element

N	WORD	FREQ.	HYPOTXT. TXT %	FREQ.	PNASX. TXT %	KEY-NESS	P
1	MAY	162	0.95	271	0.21	188.3	0
2	BE	209	1.23	677	0.53	95.6	0
3	WOULD	68	0.4	149	0.12	58.2	0
4	CELL	112	0.66	349	0.27	55.1	0
5	MIGHT	46	0.27	81	0.06	50.6	0
6	THAT	288	1.69	1,349	1.06	47	0
7	SUGGEST	32	0.19	51	0.04	38.9	0
8	SUGGESTS	29	0.17	42	0.03	38.4	0
9	LIKELY	26	0.15	35	0.03	36.6	0
10	COULD	42	0.25	101	0.08	31.7	0
11	TAND	50	0.29	136	0.11	31.3	0
12	APPEARS	24	0.14	37	0.03	30.1	0
13	IF	47	0.28	130	0.1	28.6	0
14	SVAU	50	0.29	154	0.12	25.1	0.000001
15	DEATH	39	0.23	107	0.08	24	0.000001
16	HAS	24	0.14	438	0.34	24.1	0.000001
17	HOLS	7	0.04	237	0.19	26	0
18	WAS	15	0.09	377	0.3	31.7	0

Table 45. KeyWords list for Hypo step.

Here are examples of concordance lines for *suggest** for the Hypo Step:

er, the results referred to above **suggest** that Si should be
evertheless, several arguments **suggest** that the antioxidant
lar architecture of the caveolae **suggests** the possibility that

nally, there is limited evidence **suggesting** that an autokinase
racts. IAME 6 9 Rx Evidence **suggests** that, under certain
13 7 b and preliminary results **suggest** a decrease in
. SRAM 58 2 The results **suggest**, instead, that even
wasps. TGRA 42 8 x The same **suggestion** can be made for
As noted earlier, initial studies **suggested** that the first
e identity over this region (31), **suggesting** a role in
this disease. TBEN 9 7 x This **suggests** the interesting
sed on these observations, we **suggest** that the phantom

Examples of concordance lines for *likely* for the Hypo Step:

8 9 x Other micronutrients are also **likely** to be part of our
particular transport step, others are **likely** to be common to
nst metabolismderived oxidants are **likely** to play an important
ves. TBEN 1 3 These events are **likely** to involve protein
ny other oxidative DNA lesions are **likely** to be important as
h modulation of synaptic efficacy is **likely** to contribute to
ll surface (50). TAND 20 4 Most **likely**, cholesterol helps
cal, primordial synthetase was most **likely** composed of the
issue-specific mutation rates are not **likely** explanations,
ormal digit. SVAL 14 5 xx It seems **likely**, therefore, that the

Examples for *appear** for the Hypo Step

1 oad. IAME 8 3 Rx Antioxidants **appear** to inhibit some of
uate diets (and smoking) of fathers **appear** to result in dama
Synthetic pesticide residues do not **appear** in the general
r one-third of the world population, **appear** to be the major
E 6 1 Four endogenous sources **appear** to account for
on. SRAM 25 4 Thus it would **appear** that there is a
ould presumably be very ancient, **appearing** as they do in
enhanced maintenance activities **appears** to be a reduction
gainst pathogens and tumor cells **appears** to be inhibited
Thus, the hypertriglyceridemia **appears** to be due to a
particles. TBRE 19 7 Thus, it **appears** that unregulated
Presence of the G1226 mutation **appears** to preclude the
Furthermore, the YAC system **appears** capable of cloning
ages. SCLE 14 8 xxx It therefore **appears** that a strict molec

Of the 47 instances of if in this Hypo Step, 37 (78.7%) appear in the sentence initial position. Here are example concordance lines for *if* for the Hypo Step:

vivo systems. SGOL 17 3 Also, **if** the clusters undergo
y to form a nonhereditary tumor, and **if** we take into account the
on the screen. SRAM 42 3 But **if** there were a slight error
MEK. SSHU 30 6 For example, **if** the hippocampi can be
carrier into the cell. TAND 11 2 **If** Ca2+ were to enter the
oss the membrane. TAND 18 7 **If** so, then mechanical
the paralyzed arm. SSHU 24 3 **If** substantiated, this kind
onconserved domain. IBEU 34 2 **If** we understood why
a basal value of 1.2 mM to 0.4 mM **if** the kinetic parameters
m that of modern PSI-C, particularly **if** the clusters did not
and H. mobilis RCs. SGOL 19 4 R **If** the correspondence
isual feedback. SRAM 49 5 xx Third, **if** the patient uses the
hemical action, SVAU 20 6 R Thus, **if** activation of a recep
al spatial information. TBEN 12 8 x **If** this is the case, an
sulfur bacterial RC(48). SGOL 19 5 x **If** this prediction holds,
ber of antioncogenes. HKNU 31 6 xx **If** the LCFS2 mutation

5.5.2 Move Step, synthesizing: Dev – Explaining possibilities for further development and applications

The only KeyWords are *virus* and SWIL. Here are some sample Development Step sentences.

SWIL 11:4	R	*Even a reduction in systemic virus titer* ***should help*** *control field epidemics by [...]*
TAND 0:5	xx	***Insights gained from studying*** *potocytosis* ***suggest several different ways that*** *this membrane specialization might function to integrate incoming and outgoing cellular messages.*
TAND 21:1		***The next step is to learn how*** *caveolae help the cell to receive and interpret environmental stimuli.*
SWIL 21:5	RPx	***Three possible modes of action have been proposed,*** *which could apply singly or in combination to all examples of replicase-mediated resistance described above and below: [...]*

5.5.3 Move Step, synthesizing: Op – Giving opinion (judgments of value, validity)

N	WORD	FREQ.	OPTXT.TXT %	FREQ.	PNASX.TXT %	KEYNESS	P
1	IKEY	76	0.54	238	0.19	53.5	0
2	WILL	50	0.36	118	0.09	51.2	0
3	BE	126	0.9	677	0.53	26	0
4	R	38	0.27	918	0.72	48.2	0

Table 46. KeyWords for the Opinion Step.

The Plot function showed a tendency for this Step to occur toward the end of the article, except in the case of IEPS and IKEY reviews in which instances are scattered throughout the paper. The IKEY text, which was originally presented as a speech, has a particularly high frequency of this Opinion Step. An interesting finding is the high frequency of R, or reference citations, for this step. As can be seen from the examples below, this is because these sentences are expressions of the authors' evaluation of others' work.

HKNU 17:2	R	***Especially interesting is the observation that*** normal cells, but not those mutant at TP53, are arrested temporarily in G1 by ionizing radiation (32).	Op
HNIC 14:1	R	***Interestingly***, treatment of MOLT-4 cells under appropriate conditions with enediyne 40 followed by observation of cell morphology and cell death ***revealed*** the phenomenon of programmed cell death (apoptosis) (48) ***as the prevailing cause of*** cell destruction (49).	Op/ LitH
IBEU 22:6	R	***Not surprisingly, then,*** very little of the administered enzyme is found in bone marrow cells, even immediately after marrow infusion (20).	Op/ Phe
SGOL 7:1	R	***One surprising finding is that there is no evidence for*** a second RC gene in C. limicola that would code for a second homologous RC protein (10) (***the data will be available shortly for*** H. mobilis).	Op/ Gap

SRAM 45:1	R	***Obviously***, MEG ***studies could also be useful in*** exploring remapping ***effects of the kind observed by*** Pons et al. (33) and ***our group has recently undertaken such a study*** (49) ***in collaboration with*** Christopher Gallen and Tony Yang.	Op/ CR
SWIL 33:6	R	***Paradoxically, the first demonstration of*** a natural DI molecule attenuating virus disease symptoms in a transformed plant involved a defective, subgenomic ssDNA of the B component of African cassava mosaic geminivirus (108,109), ..	Op/ His

5.6 Summary of Step features

Summarized below are the features of the Steps described above. The Discourse Steps are global items functioning mainly to organize the text at the whole text, section and paragraph levels. However, they can appear in the same clause (phrase, as in headings and rare instances of isolated clauses) as a Move Step, which is used to present a Move. The Moves are discourse community-related information items which are used to express the author's intent or information which the author wishes to convey. The Discourse and Move Steps are realized by clause elements, such as SIE, Vb, and key words and phrases.

Step: Topic
Function: Introducing a topic
Notes: *Almost always first sentence of a paragraph/section
*Often combined with Move Steps Lit, Phe, Gap, AsFL, His
*Usually evenly distributed throughout text

Step: Aim
Function: Expressing aim
Notes: *Appears early in paper
*Often appears at end of paragraph/section
*Often uses present tense verb, *will, shall*
*Frequent use of Dmacro/Dmicro discourse function of grammatical subject

Key Words

review	*This review summarizes*
this	*this review, this paper, it is the purpose of this article to, in this essay*
discuss	*we/I discuss*

Step: Disc
Function: Organizing discourse – referring
Notes:
*Frequent use of present tense verbs
*Use of imperative sentences and questions
*Often appears at beginning or end of a paragraph
*Use of imperative sentences and questions
*Use of *we* to include audience in discussion
*Used with other Move Steps: AsFL, Hypo, Op, Sum
*Used in 19/25 texts usually scattered throughout text when used

Key Words

what	*what would happen if*
table here	*Table X shows/lists/depict/is based on, shown in Table X, listed in Table X*

Step: Sum
Function: Summarizing
Notes:
*Frequent use of present tense verbs
*Frequently found at end of a paragraph
*Often found with other Move Steps, e. g. Op

Key Words

thus	Most frequently at beginning of sentence; often in the sentence ending a paragraph

Step: Def
Function: Giving a definition
Notes:
*Uses copular in present tense with RWE GS for SIE
*Frequently the first sentence in a paragraph
*Most often found in Theory and Status quo reviews

Step: His
Function: Describing historical events
Notes: *Most frequently found in History reviews
*Frequent use of past tense
*Frequently with reference citation
*Scattered in text in History reviews, toward first half of articles in Status quo and Theory reviews

Key Words
first Use with verbs, e.g., *first appeared;* use with nouns, *the first member of , the first synthesis of*

Step: Phe
Function: Describing generally accepted phenomena
Notes: *Most frequently found Step
*Significantly high frequency in TGRA, IAME, IKEY, SCLE, SSHU, SRAM, HOLS
*Present tense verb
*Frequently with reference citation
*Predominant use of RWE GS for SIE

Key Words
we Used frequently in Issue reviews

Step: Tech
Function: Describing key technology, method, tool
Notes: *Most frequently used by SSHU, HOLS, SCLE
*Frequent use of present tense verbs
*Numerous reference citations

Key Words
method *a convenient/simple/heuristic method; this method;* Name *method*

Step: CR
Function: Describing current research – not specific, general and current

Notes: Most instances in SRAM
Sometimes with reference to unpublished material or personal communication
Use of present continuous tense

Key Words
recent *recent evidence/studies*

Step: Ela
Function: Presenting examples, explanation
Notes: Sometimes with reference citation

Key Words
example *for example* – often sentence initial position

Step: Lit
Function: Introducing related literature

Key Words
show Verb: present perfect >> past >> past perfect >> present; *was/were shown to* V; *has/had shown that; has/have been shown to* V
recently *Recently, More/Most recently, has/have recently* V-*ed*

Step: LitH
Function: Introducing historical literature
Notes: In History reviews
Author's name mentioned in citation

Step: HisH
Function: Hypothesizing about historical events
Notes: Used in most frequently in HNIC

Key Words
would *would require*

Step: AsFL
Function: Assessing findings, literature
Notes: Citation author name usually given for complimentary reference or for specific contradiction
Often associated with other Steps: Dev, His, Hypo, Lit, LitH, Op, Phe, Sum

Key Words

these	*these data/experiments/observations/results;* frequently in sentence initial position

Step: Gap
Function: Describing a gap or limitation in the knowledge of the field
Notes: Other frequently used negative signaling words include: *no, but, however, although* and *unfortunately*.
Shows tendency to cluster

Key Words

not	*not apparent, has/have not V-ed, it is not clear, we do not / it is not known, not well understood, not yet V-ed*
remain	*remains to be V-ed*
whether	*determin* whether, it is not known whether, question of whether*
no	*no evidence (for)*
difficult	*it is difficult to V, is/are difficult*

Step: Hypo
Function: Hypothesizing – about specific phenomena: trying to explain

Key Words

*suggest**	*These data suggest, limited/preliminary/a wealth of evidence suggests that, These findings suggest, These observations suggest that, results suggest that, strongly suggests, studies suggested, This suggests (that), we suggest*
likely	*is/are likely to V, most likely, It seems likely that*
appear	Real World Entity subject *appear to* V, *it appears that*
if	Very often used at (78.72%) as sentence initial element

Step: Dev
Function: Explaining possibilities for further development and applications
Notes: Mainly used in SWIL

Step: Op
Function: Giving opinion – expression of judgments of value, validity
Notes: High frequency of reference citation
Appears in all reviews, but most often in Issue reviews
Tendency to cluster at end
Prominently used throughout IBEU, IEPS, IKEY, SCLE and SVAL

Both the Discourse and Move Steps are used in various combinations to compose a review text describing the history of a discourse community field of interest, presenting the status quo, proposing a theory or model, or drawing attention to an issue in the field. The next chapter will examine how these Steps cluster into Moves which compose the different types of reviews.

6. Review types, moves and Steps

Title, abstract and subheadings make the content clear.
(questionnaire response to question on decision factor about review article purpose, Neil Thomas, March 5, 1996)

6.1 Review type classification

The reviews were classified into four types based on information from the journal instructions, specialist informant responses to questionnaires, and examination of the review texts themselves. Let us now try to collate the discussions on reviews, i.e., the review types, moves and steps. As noted by Thomas, in the quote above, the aim of the review can often be grasped from its "title, abstract and subheadings". Reviews do not have the familiar IMRD (introduction, materials and methods, results and discussion) format of research articles, and therefore headings, where used, are indicative of the content itself. Presented below are the headings for each of the reviews together with the steps, their features, and notes on author intentions, some of which were commented on by the specialist informants.

6.1.1 History Reviews

The 25 reviews examined included four which were classified as history reviews:

HBRO	A brief history of opiates, opioid peptides, and opioid receptors
HKNU	Antioncogenes and human cancer.
HNIC	Chemistry and biology of natural and designed enediynes
HOLS	The human genome project

These history reviews tended to have a chronological description of events (research reports) in the field and a much shorter description of current position and projection about future work, often only in the last few paragraphs.

HBRO, as suggested by the title, was the shortest with 21 paragraphs and 2,237 running words. This review had no abstract or internal headings. In place of headings, the author uses many TRWE (time real-world entity) signals as the SIE to guide the reader from the prehistory introduction through to the description of modern research in the field. It starts with a quote from The *Odyssey* by Homer to catch the readers' attention and make them aware that opium and opiates have a very long history, with *opos* coming from the Greek term for *juice*. The description of ancient history continues from paragraph 1 through 5, followed by accounts of these compounds in the 1800s (paragraphs 6 and 7), after 1939 (paragraph 8), after 1942 (paragraph 9), in the 1960s (paragraph 10), from 1975 (paragraph 11), and from 1981 (paragraph 12). TRWE signals such as "by the mid-1960s" and past tense verbs are used to guide the reader. From paragraph 6, the narrative changes to a description of scientific research into this class of compounds, starting with:

> In 1806, Serturner (8, 9) isolated the active ingredient in opium and named it morphine after the god of dreams, Morpheus.

From paragraph 13, the description is of current knowledge and research. Paragraph 20 describes very recent research:

> *Until this year, the structure of opioid receptors remained a mystery. Now two groups of investigators have published descriptions of* the expression cloning of cDNAs encoding the o receptor on the neuroblastoma-glioma (NG108-15) cells (36, 37).

Paragraph 21 concludes the review on a very positive note on the cloning of a new receptor:

> *Surely a new chapter in the annals of opiate research is about to be written.*

The chronological flow of this review is also obvious from the Plot function of the Wordsmith tools which visually presents the location of word occurrences within a text. Doing a plot of the Keywords shows that the *opium* appears frequently in the first one-third of the text followed by the word *opiate* for the next one-third and then by *opioid* for the final one-third. Also, the words *receptor* and *receptors* appear only in the second half of the text. (The Keywords plot could not be reproduced here.)

The other three history reviews were longer, ranging from the 4,699 words of HNIC to the 7,351 words of HKNU (see Chapter 3, Tables 10 to 17), and were divided into sections with headings. HKNU, with a total of 40 paragraphs, begins by stating that recent discoveries linking cancer and genetic predisposition can be promising. The last sentence in the fourth paragraph wraps up the introductory section with

> *These discoveries have illuminated our understanding of* human cancer *considerably* and *have revealed* a group of genes *that is important in* cellular and developmental biology, *making this a propitious time to review progress on the subject and to take note of some future prospects for extending this understanding.*

The review ends with a section on 'The search for new antioncogenes', starting with paragraph 38, which makes predictions and projections about future work.

The HNIC review on enediynes, which are used as anticancer agents, is composed of 25 paragraphs divided into five sections. The authors begin the main text by stating:

> *Nature continually serves to provide scientists with new ideas, often provoking us to reexamine previous studies and leading us in new directions.*

They then draw attention to the enediyne unit by stating:

> *For at the very heart of these molecules is* contained the same enediyne unit; and *therein lies the means by which* these unprecedented molecules exert their *remarkable* biological properties.

After describing the knowledge base in detail, they end with a section on 'Concluding remarks and future prospects' which runs from paragraph 23 to 25.

The HOLS review, composed of 45 paragraphs divided into six main sections, specifically describes the Human Genome Project in the United States. It tries to justify its aims, define terminologies, describe goals and limitations, summarize developments, and explore further avenues for development. The introductory paragraph ends with

> After a 5-year reality check, *it is of both scientific and policy interest to examine how* the committee's view toward this project has fared in the field.

The final paragraph is a positive one with statements such as:

> Like other human endeavors, the Human Genome Project *has succeeded best when it has aligned itself with broader trends.*

The note added in proof reports the identification of a gene described in the report, thus attesting to the rapid progress in the field.

In sum, the history reviews seem to use the following moves and move steps.

Moves	Move Steps
Describe (chronological) background	His, Def, Phe, Lit
Describe current position	Phe, Tech, Ela
Point out issues	AsFL, Gap, Op
Suggest future work	Hypo, Op

6.1.2 Status quo reviews

Nine of the 25 reviews were status quo reviews describing the current status of the field.

SCLE	Chloroplast gene sequences and the study of plant evolution.
SGOL	Shared thematic elements in photochemical reaction centers.
SLAI	Hepatocyte nuclear factor 3/fork head or "winged helix" proteins: A family of transcription factors of diverse biologic function.
SLEN	Tumor-suppressor genes: News about the interferon connection.
SRAM	Behavioral and magnetoencephalo-graphic correlates of plasticity in the adult human brain.

SSHU Nuclear magnetic resonance imaging and spectroscopy of human brain function.

SVAL Molecular analysis of the microtubule motor dynein.

SVAU Toward an understanding of the molecular mechanisms of physiological cell death.

SWIL Strategies to protect crop plants against viruses: Pathogen-derived resistance blossoms.

The SCLE review very clearly states its aim to be as follows:

> *A large cooperative effort has emerged to produce* a detailed phylogeny of the seed plants in general, and flowering plants (angiosperms) *in particular,* based on sequences of the chloroplast-encoded large subunit of ribulose-1,5-bisphosphate carboxylase (rbcL).
> *The purpose of this review is to give a brief account of the history and status of this effort.*

It thus goes into detailed descriptions of the phenomena involved, including explanation of theory, methodology, assessment, and limitations, and then closing with suggestions about further possibilities and developments. The author uses sentence headings to introduce the main topics and uses cycles of presenting and evaluating move steps with evaluating steps toward the end of each section. The word *data* appears as a keyword when this review is compared with others in the Status quo review database. The plot for this word shows that *data* appears from the beginning of the text and is distributed throughout except for its absence in the few introductory paragraphs (Para 2-6). The final paragraph of the review follows the heading *Conclusions* in which the author returns to emphasizing the extensive body of data now available and ends with the need to address "many other questions".

SGOL, in the abstract, presents the aim of the review as follows:

> *This review summarizes the current state of knowledge in* comparative reaction center biochemistry between prokaryotic bacteria, cyanobacteria, and green plants.

This is restated at the end of paragraph 2 before the author launches into a description of what is known and then presents and assesses various hypotheses. The last three paragraphs contain many author opinion statements based on analyses of current developments.

SLAI is a multi-author review which starts with a description of historical background, recent work in the field, and a statement of aim at the end of paragraph 3:

> *In this review we will summarize the recent biochemical, structural, and biologic information about* this protein family.

This is followed by an assessment of available information and unsolved issues, a description of an analysis method and the phenomena observed, and then a description of current work and issues. Paragraph 11 summarizes the paper and suggests that rapid progress can be expected using the protein-DNA complex described.

SLEN presents the history and phenomena related to tumor-suppressor genes in the first three paragraphs before discussing their activity. This is followed by cycles of steps giving definition, literature presentation, gap and hypothesis. The author then summarizes with an assessment and synthesis of the information presented, ending as follows:

> *A commentary from 1990 entitled* "Interferons, a New Class of Tumor Suppressor Genes?" *carries a question mark in its title* (37). *Were we to consider* the interferon genes together with the genes whose expression they control, *then the recent findings appear to warrant that now the question mark be replaced by a period.*

The SRAM paper was the longest, with 8,390 running words and 59 paragraphs. It was also particularly interesting because of the copious amount of current research that it contained. The author Ramachandran clearly states at the beginning his intention to do this:

SRAM 7:1		*My purpose in this essay is to review the evidence for* somatosensory plasticity in the adult mammalian brain.
SRAM 7:2	R	*Some excellent reviews are already available on* the physiological work (13-16) and I will therefore *mention* these experiments *only briefly*.
SRAM 7:3	xxxx	I *will focus, instead, on our own* behavioral *work* and on the MEG correlates of somatosensory reorganization in humans.

Yet, at the end, the author presents a challenge to the status quo, suggesting that he may actually have intended to write an Issue review. In fact, in 1998, Ramachandran (with Blakeslee) published a popular science book, “Phantoms in the Brain,” which was received with much acclaim and was selected as a finalist in the 1998 Los Angeles Times Book Prizes. In this book, he states that he began his research into brain plasticity in 1991 (Ramachandran and Blakeslee 1998: 25) partly because the medical textbooks he had consulted offered no explanation for them. Therefore, in retrospect, the 1993 review might be thought of as a challenge to the community in this field of study. The implications of presenting ideas in review articles, textbooks and popular science textbooks are discussed further below.

SRAM 58:1 *Taken collectively, the findings on* visual and somatosensory filling in *considered in this review challenge* two of the accepted dogmas in neuroscience: the concept of the receptive field as a set of receptors simply converging onto sensory neurons and the concept of fixed maps or topography.

SRAM 58:2 *The results suggest, instead, that* even in adults, the classical receptive field is just the tip of the iceberg – its profile can be altered by ongoing visual stimulation in the surround (as in artificial scotomas).

The SSHU review clearly indicates in the title and abstract that this review will discuss the use of these methods for study of the human brain function. The first paragraph presents the historical background, the second and third describe recent technological advances and the fourth states the aim of the review.

> *It is the purpose of this article to show how recent developments of* MRS and MRI methods for studying the human brain *are beginning to provide information about* the spatial localization and metabolic changes associated with human brain activity.

A description of the technologies follows with descriptions of studies using them and the last three paragraphs, following the heading “Future Developments in Functional MRI/MRS”, project about future field developments.

SVAL presents background information related to dyneins and the reasons for its importance with a clear statement that work on these polypeptides will be reviewed in this paper. The author then uses cycles of presentation of current work in the field followed by assessment and opinions of them. The author concludes with the claim that continuing work "should be of considerable interest" and adds a note in proof describing a recent achievement from another laboratory, suggesting much activity in this field.

The SVAU review is trying to develop a way to understand the molecular mechanisms of physiological cell death. There are many cycles of statements describing phenomena observed, assessment of findings, and hypothesizing. What seems to be like a groping about is due to the fact that the field, at the time, was in the process of very active development, as evident from the ending statement:

> *This search is just now beginning, especially through the application of* molecular biological *techniques* and genetic *experiments in well-defined model systems such as* C. elegans.

The SWIL paper begins with a clear statement of the aim in the abstract:

> *This review attempts to consider, in a critical way, the current state of the art, some exceptions, and some proposed rules.*

The review itself starts immediately with "Some recent history", introducing the subject with background literature and the importance of this field. The author then starts a cycle of phenomenon explanation, assessment of available data and suggestions for future exploitation. The last two sentences close with strong author opinion statements:

> *In conclusion, I am confident that we shall continue to be amused and amazed by the ingenuity and unpredictability of future strategies for* pathogen-derived resistance to viruses. *Let us hope that all this effort, new information, and experimental material also contribute to our knowledge and understanding of* conventional virus-plant interactions.

The moves for the status quo reviews can be summarized as follows:

Moves	Move Steps
Present background	Def, Phe, Lit, LitH, His, HisH
Present phenomena	Phe, Lit, Def, CR
Present important technologies	Tech, Lit, Ela
Present issues	AsFL, Gap, Op
Offer hypotheses, explanations	Hypo, Op
Suggest future work	Hypo, Op

These reviews often start with some historical background of the topic but quickly move into descriptions of phenomena and technologies. The Wordsmith Plot for the History Step shows that SGOL, SLAI, SLEN, SRAM, SSHU, and SVAU have instances of this step clustering within the first one-fourth of the paper. After that there are mostly cycles of introduction of phenomena, commenting on it, and expressing hypotheses or opinions about it.

6.1.3 Theory/model/hypothesis reviews

Six of the 25 reviews were papers that aimed at proposing or describing a theory, model or hypothesis.

TAND	Caveolae: Where incoming and outgoing messengers meet.
TBEN	The molecular machinery for secretion is conserved from yeast to neurons.
TBRE	Transgenic mouse models of lipoprotein metabolism and atherosclerosis.
TGRA	Modes and origins of mechanical and ethological isolation in angiosperms.
TPRI	Alzheimer disease and the prion disorders: Amyloid â-protein and prion protein amyloidoses.
TSCH	An operational RNA code for amino acids and possible relationship to genetic code.

TAND started with a description of phenomena, issues and possible hypotheses, and then presented the aim of this review as outlining the data and using this information to construct several models for this topic. This is followed by cycles of definition and literature reports with their assessment and presentation of hypotheses. The author then presents his own theory on how caveolae function and concludes in an "Outlook" section in which he states:

> *The next step is to learn how* caveolae help the cell to receive and interpret environmental stimuli.

The co-authored TBEN review proposes "a model for synaptic vesicle docking and fusion" by focusing on "several integral membrane proteins". There are cycles of historical or phenomenon description followed by presentation of gaps in the knowledge and hypothetical statements. To soften the conclusion, the last paragraph includes a disclaimer:

> *We do not intend to imply that when we understand* membrane flow in yeast *that we will completely understand* the molecular basis of synaptic transmission.

The TBRE review describes the key technology which is the principal subject of the review, i. e. transgenic mice. It starts with a general introduction to the overall topic, some definitions of terms, and an explanation of the relevance of the topic to human disease. There is next an introduction to previous studies, the key genes which have been manipulated in the transgenic mice, and the control of their expression by gene and nutritional regulation. After a general introduction to manipulation of genes in transgenic mice, there is a description of specific examples and reviews results. This is followed by cycles of model description, assessment, opinion and concluding statements. Finally, the new topic of using transgenic mice for the study of a particular disease is introduced and some findings are described. The concluding statement is:

> *These* animals *should be of great assistance in studies of* diets, genes, and drugs influencing atherosclerosis and *should greatly accelerate research in this field.*

As a keyword in comparison with the Theory review corpus, the word *mice* ranks first and the Plot function of Wordsmith shows that it appears throughout the paper. The other interesting Plot data is the word *model* which appeared as a keyword when this review was compared with the total review corpus, showing that this word appears more commonly in Theory reviews than in the other types. This word also appeared 14 times scattered rather evenly throughout the review.

The TGRA review begins with a review of the literature then moves into cycles of definition, hypothesis and phenomena description to fulfill its two aims, one stated in paragraph 4 and the other in paragraph 6:

> *One of the purposes of this paper is to present the broader data base that now exists.*

> *Thus this paper will examine* the diverse origins of floral isolation.

The ending raises many questions that remain for further work.

The multi-author TPRI review begins with a clear explanation in the abstract that a new perspective will be presented on Alzheimer disease and the prion disorders. The main text itself describes the topic with the related literature, assessing the information and presenting the author's hypotheses and opinions. The final paragraph states:

> *In conclusion*, AD and the prion disorders *show many parallels*. *Ongoing* genetic, neuropathologic, and cellular/molecular biological *investigations of* AD and the prion diseases *will enormously enhance our knowledge of the mechanisms of* these two types of neurodegenerative disorders.

The TSCH review, written by four authors, has the following two aims:

> Seven *examples of* sequence-specific aminoacylation of RNA helices based on tRNA acceptor stems *are described here*.
> *We also discuss what is known about* the minimal structures in synthetases, which are responsible for aminoacylation of these substrates.

The review starts with definitions and phenomena descriptions followed by a discussion of problems. This is followed by a consideration of the work in the literature leading to the concluding statement of

> *These experimental facts provide a specific framework within which theories for* evolution of the genetic code *can be constructed.*

The relationship of moves and move steps for Theory reviews is summarized below. Theory reviews try to propose or describe a theory, model or hypothesis for some phenomena in the field of study. As can be seen from the table and the TBRE analysis below, there are cycles of presenting, evaluating and synthesizing steps, with, as noted in Chapter 5, Section 5.3.3, the highest average percentages for the Phenomenon Step and a relatively higher percentage for the Hypothesize Step.

Moves	Move Steps
Present background	Phe, Def, Lit, Gap
Present phenomena	Phe, Lit, Tech
Describe theory/model	Hypo, Op, AsFL
Support proposal	CR, Lit

6.1.4 Issue reviews

The six issue reviews were the following:

IAME	Oxidants, antioxidants, and the degenerative diseases of aging.
IBEU	Gaucher disease as a paradigm of current issues regarding single gene mutations of humans.
IBLE	Signal transduction via the MAP kinases: Proceed at your own RSK.
IEPS	The anomaly of silicon in plant biology.
IKEY	Are there ecological limits to population?
ITEM	Retrovirus variation and reverse transcription: Abnormal strand transfers result in retrovirus genetic variation.

The IAME presents an argument to encourage Americans to consume more fruits and vegetables to improve health by presenting the reason for aging and degenerative diseases:

> *We argue that* this damage (the same as that produced by radiation) is a major contributor to aging and to degenerative diseases of aging ***such as*** cancer, cardiovascular disease, immune system decline, brain dysfunction, and cataracts.

It is composed of cycles of topic presentation, phenomenon and literature explanation, assessment and author opinion. From paragraph 14, the focus shifts to the argument for antioxidants with cycles of topic and literature presentation followed by assessment and authors' hypotheses and opinions. The two concluding paragraphs call for more consumption of fruits and vegetables to increase intake of antioxidants and thus improve health.

The title of the IBEU review presents the author's viewpoint that Gaucher disease can be used as a paradigm for the consideration of other serious diseases related to genetic diseases. Paragraphs 1 to 3 present the overall introduction and aim of the paper, which has self-explanatory headings to guide the reader. Historical background is presented, followed by cycles of topic presentation and phenomena description, with assessment of situation and author opinion. There is also a presentation of ethical issues related to genetic diseases and here, too, the author presents his opinion. The review ends with a repetition of the aim to offer this paradigm for the consideration of the issues related to other genetic diseases.

The IBLE review begins by examining the literature:

> *A flurry of reports in the last year demonstrated just that.*
> *Here I discuss some of these studies, with their various caveats, that provide new information aiding in the further* biochemical and molecular *dissection of* the complex signaling systems regulating MAP kinases and RSK.

It then presents the topic and the phenomena, followed by assessment and hypotheses. Finally, the aim of the review is repeated to call attention to risks involved in trying to describe the complex signaling network:

> *As indicated by the title of this review, there is great risk in trying to describe* the ever-expanding, complex signaling network involved in the regulation of MAP kinases.

The IEPS review tries to point to the importance of silicon in solution cultures used to examine plant behavior:

> Omission of silicon from solution cultures *may lead to distorted results in experiments on* inorganic plant nutrition, growth and development, and responses to environmental stress.

After the abstract is a quote from the work *Principia Mathematica*:

> It is to be observed that a definition is, strictly speaking, no part of the subject in which it occurs. (Alfred North Whitehead and Bertrand Russell, *Principia Mathematica*)

The review itself is composed of cycles of phenomenon description with historical and literature references followed by author assessment and opinion presentation. The conclusion restates the importance of silicon in other fields, including the possibility of "taxonomic, anatomical, archaeological, and medical implications" and recommends greater awareness of this mineral.

The IKEY review was originally a presentation given at the 130th Annual Meeting of the National Academy of Sciences on April 27, 1993. The author, a renowned scholar in demography and sociology, calls for clearer policy on population growth based on biological and economical concerns. In this review, he summarizes the differences in methods and traditions of economics and biology as eight axes, presenting the topic with numerous references and quotes to support his points. He ends on a modest note:

> It would be impertinent for me, a professional in neither field, to inform economists and biologists on their disciplines. That is not what I have tried to do. My effort has been rather to say how these two disciplines look from the outside and still from the outside, which of their features *account for* the different stands they take on population. I do not *interpret the differences between* them as errors but speak *rather* of different problem formulations, arising out of the different ways that the disciplines serve the cause of knowledge and the welfare of society.

This ITEM review is unusual in that it includes 19 references, scattered throughout the text from Para. 4 to 24 (last Para 27), to unpublished material with the author named with another researcher (the entire corpus of 25 reviews has only 34, with the remaining 15 found in seven different texts). As stated by informant Wood, this seemed unusual for a review because the author seemed to be presenting many of his "own views and data". Perhaps this was accepted because the author is a leading specialist on retroviruses, being one of their discoverers. He tries to explain why retroviruses, which are related to diseases such as AIDS, display numerous mistakes in reverse transcription. The argument is based on six points, which are presented in cycles of statement of the component of the problem, followed by supporting evidence and description of related aspects.

The moves and move steps for the Issue reviews are summarized below. The text evidence for this group of reviews points to the discussion of some issue or problem in the field. Like the other types of reviews, background information is presented and phenomena and technology are described, but the focus is on some problem which is presented and discussed. Perhaps because of this specific focus, Issue reviews have few examples of the Define Step.

Moves	Move Steps
Present phenomenon/technology	Def, Phe, Lit, Tech, Ela
Present problem/issue	AsFL, Gap, Op
Offer solution	Hypo, Op, CR, Ela, Lit
Support proposal	Lit, CR, Op

6.2 Components of review articles

The discussion above and in previous chapters indicate that the components of review articles can be classified as shown in the list below, which also presents the supporting evidence.

Rhetorical organization: Beginning, middle, end
Evidence: Rhetorical principles, Text information
Physical features: Title, affiliation, abstract, quote, main body sections, summary/conclusion, note in proof, acknowledgements, references
Evidence: Text information

Discourse steps: Topic, Disc, Sum, Aim
Evidence: Text information, Specialist informants, Repeated use of similar sentence elements

Review types: History, Status Quo, Theory, Issue
Evidence: Specialist informants, Text information

Moves: Describe current position
Discuss problems encountered
Offer hypotheses
Offer possible solution
Present background
Present important technologies
Present phenomena
Present phenomena and technologies
Present problem/issue
Present theory/model/hypothesis
Suggest future work
Support proposal
Evidence: Specialist informants, Text information

Move Steps: AsFL, CR, Def, Dev, Gap, His, HisH, Hypo, Imp, Lit, LitH, Op, Phe, Tech, Theory
Evidence: Repeated use of similar sentence elements, Text information

To more clearly grasp the differences among the types of review articles, the moves and move steps described in the above sections for each review type are summarized in the following table. One caveat is that, as can be seen from Table 47, there is much overlap of moves and steps, and therefore, the classifications represent more 'focus' tendencies rather than rigid 'type' characteristics.

As one of the three aims of this book is to present practical ways to help nonnative English speakers with professional academic writing, in this case, the genre of the review article, let us next consider how the features of the moves and move steps presented above can be utilized to create useful writing tools.

Review focus type	Moves	Move Steps
History	Describe (chronological) background Describe current position Point out issues Suggest future work	His, Def, Phe, Lit Phe, Tech, Ela AsFL, Gap, Op Hypo, Op
Status quo	Present background Present phenomena Present important technologies Present issues Offer hypotheses, explanations	Def, Phe, Lit, LitH, His, HisH Phe, Lit, Def, CR Tech, Lit, Ela AsFL, Gap, Op Hypo, Op
Theory	Present background Present phenomena Describe theory/model Support proposal	Phe, Def, Lit, Gap Phe, Lit, Tech Hypo, Op, AsFL CR, Lit
Issue	Present phenomenon/ technology Present problem/issue Offer solution Support proposal	Def, Phe, Lit, Tech, Ela AsFL, Gap, Op Hypo, Op, CR, Ela, Lit Lit, CR, Op

Table 47. Moves and move steps for the different review focus types.

7. Writing-support concepts and tools

> According to Vygotsky (1987), tools mediate human effort to affect the physical world. That is, tools are used to increase our mechanical, chemical, electrical, and other types of advantage in such a way that we can accomplish tasks not otherwise possible.
>
> (Cunningham et al. 1993:43)

7.1 'Publish or perish'

'Publish or perish,' a much maligned adage, does contain a kernel of truth in the academic world and particularly in the sciences where there is sometimes heated competition for being the first to publish a finding. In a professional discourse community, writing is an essential means of gaining a voice. However, because the predominant language of international publication today is English, this can disadvantage those who do not have the 'tools of English'. The 'tools' refer to a working knowledge of the language which can be applied at will and with skill to express the writer's thoughts and intentions. These tools therefore would not necessarily have to be mechanical or computer-related entities but could include frameworks and approaches to resolving issues in order to fulfill a purpose. Various 'nested issues' can obstruct efficient and effective expression when working in a language that is not one's first. To help overcome these problems, let us now examine ideas from various fields, including computer technology and robot engineering, to find ways in which external tools, as opposed to internalized native-language ability, might be developed to support efforts by NNS (nonnative speakers of English) to publish for their professional discourse community.

The use of writing-support tools would not compromise the writing process but rather serve to free the writer to focus on optimally packaging the information to be delivered, as suggested by the quote above by Cunningham et al. (1993: 43) after Vygotsky (1987). This idea is also expressed by Kanaoka (1999), who in describing a technical writing course for designing effective technical documents, refers to a call by Pellegrino (1995) to teach students how to take advantage of multimedia tools. Kanaoka (1999: 19) points out that "technology lets us focus on the logic of what we are doing rather than keep track of all the details. Our thought, in both memory capacity and its conscious manipulation, is severely limited." What is relevant to the present discussion is that working in a second language could limit this even more but this disadvantage could be reduced with effective writing-support tools.

This book has described the various issues of professional discourse as being 'nested' issues, which were categorized into three types dealing with basic grammatical and technical issues, word/ phrase and sentence level issues, and paragraph/text issues. The complexity of these issues partly arises from the fact that they involve the expectations, not necessarily 'explicitly' expressed, of three different 'communities': the English-speaking community (ESC), the English-speaking academic community (ESAC), and the focused global discourse community (FGDC). The present chapter will propose 'tools,' which might help resolve some of these issues, by examining work being done in information science, computer science and educational technology.

The term 'writing-support' is being used in preference to 'style manual' or 'automatic translation'. This symbolizes the importance of NNS scientists retaining the freedom to express their ideas with their own rhetoric. This is done with the hope that effective support can be provided without stifling the possibility for the presentation of new vistas arising from different worldviews. Whether or not this is possible is to be debated, but as discussed when considering cross-cultural issues, it would be of value to both the individual and the FGDC to entertain the possibilities offered by the expression of genre characteristics using alternative rhetorics. Without this freedom, we would not have the evolution that can occur at the edge of chaos.

7.2 Towards writing support at the lexical level

NNS writers need support at grammatical, lexical and rhetorical levels. As will be pointed out below, the major problems lie with the last two, which will be discussed in turn in the following sections.

Genre analysis has done much to identify the FGDC expectations and thus aid the NNS trying to write research papers in English (Weissberg and Buker 1990, Huckin and Olsen 1991, Swales and Feak 1994). However, even more help is needed, as noted by Gosden (1992b) who found that the editors of more than a hundred academic science journals in North America and the U.K. acknowledge that the NNS faces linguistically related problems in addition to the usual difficulties encountered with FGDC publishing. Gosden (1996), working with Japanese scientists at a technological university, also interviewed his students who had to publish in English in order to successfully complete their doctoral program. He found that that many "write with others' words", a phrase which he quotes from Campbell (1990). One of Gosden's (1996: 118) informants says: "I choose famous authors and address [sic] [...] I underline some good points and use the vocabulary, patterns, for example 'proceeded cleanly', 'there has been much interest', or Figure captions. I underline such points and use them when I'm writing".

This "write with others' words" may smack of plagiarism. Of course, this is certainly not what is being recommended here. However, the need for help at the phrase level was ingeniously shown by a study by Okamura and Shaw (2000) on the differences among transactional letters written by academic professionals and non-professionals who were native speakers of English and nonnative speakers of English. They use the term "subculture" for what I refer to in this book as 'discourse community' and the term "international diaculture", or a global subculture of scientists, for my FGDC concept. Okamura and Shaw (2000) had the four groups of NS and NNS who were or were not members of their international diaculture write a cover letter to the editor of an international journal to accompany the submission of a manuscript. They classified their findings into the three categories of grammatical, rhetorical and lexical and, noting that all groups displayed acceptable grammatical performance, showed that the rhetorical awareness important to genre texts was lacking among nonprofessionals whether or not they were native speakers of English. Their other important finding was the importance of teaching lexical phrases to the NNS as these were "signals of insider status" (a concept from Nattinger and DeCarrico 1992). Table 48 was prepared from the information in the Okamura and Shaw paper (2000).

	NS-prof	NNS-prof	NS-nonprof	NNS-nonprof
Grammatical	O	O	O	O
Rhetorical	O	O	X	X
Lexical	O	*	*	X

Table 48. Appropriateness of text features used by different groups. (NS = native English speaker; NNS = nonnative English speaker; prof = professional; nonprof = nonprofessional; O = appropriate, * = partially appropriate, X = inappropriate)

Problems with vocabulary have been recognized for a long time. Robinson (1980: 71) quotes Fanning (1977) "that it is not single words which are always difficult, but phrases, so that common combinations of words should be taught, not just the individual vocabulary items of the discipline." However, being able to offer timely and individually tailored help has only recently become feasible with developments in corpus linguistics and also in computer technology, which now offer various support tools for language study and production.

One caveat is that this technology can be both advantageous and disadvantageous. The fact that we now have the tools to study language quantitatively is a great advantage but this can also lead to too rigid an approach to language if there is too much reliance on the usage 'rules' that become evident through such research. The editorial in the inaugural issue of *International Journal of Corpus Linguistics* (Teubert 1996) notes that corpora had been around for about thirty years at that time and had already been changing our view of language. It also boldly claims that "language engineering is the core of information technology, and information technology will be the key industry of the 21st century" but, more importantly recognizes that "a language without fuzziness, opacity, and metaphoric capacity brings about a petrification of ideas" (Teubert 1996: ix).

In sum, there is a need to identify lexical patterns of text to make them accessible to NNS and yet respect the "fuzziness, opacity and metaphoric capacity" of language. Trying to do both may seem to be self contradictory but a possible solution is to offer as much freedom as possible without violating the expectations of the discourse communities involved.

7.3 Towards writing support on the genre text level

This study was begun to explore ways to aid the NNS writer from the viewpoint of second-language pedagogy. Most of the available materials have been developed from a 'narrow' point of view of language pedagogy alone. What was needed was more input and synthesis of concepts from the natural and engineering sciences where research is being extended from the study of the natural external world, including medicine which deals with the physical aspects of human beings, toward the workings of the human mind and language. *Scientific American*, in its 150th anniversary issue on the 'Key technologies for the 21st century', includes technologies related to artificial intelligence, intelligent software, and virtual reality. In the commentary on artificial intelligence, Lenat (1995: 64) predicts the development of "programs that understand natural languages" which can use knowledge bases to comprehend texts laden with the ambiguity that plagues machine translation. He predicts the advent of word processors that "will check content, not just spelling and grammar" (Lenat 1995: 64).

Within two years after Lenat's article, much progress had been made in the artificial intelligence field. At the March 1997 conference in Silicon Valley convened by the Association for Computing Machinery, historian James Burke asserted that "we stand today on the threshold of an explosion in information technology, the social and economic consequences of which will make everything that came before look like slow motion" (quoted in Gibbs 1997: 64). However, before that happens, we seem to be caught today in a predicament of finding it more difficult to implement these 'advanced' technologies than they might be worth – as will be attested to by anyone who has spent hours, sometimes days, trying to upgrade an application or connect to a network or who has experienced a system crash. In fact, the phenomenal rise in labor productivity between 1960 and 1973 presents a great contrast with the increment in the period from 1973 to 1995 that reached only about one-third of increase observed during the previous period. One reason for this difference lies in the fact that the first stage of development tackled highly routine tasks, which were relatively easy to automate. Onc of the promising ways presented which may help overcome this problem is the development of specialized software – in other words, actually having the technology do less than it can rather than try to have one program do everything. A combination of useful small programs coordinated by interaction between the system and the user might offer a solution to this problem. It is interesting that the call for papers for the seventh Machine Translation Summit, held at the National University of Singapore from September 13 to 17, 1999, includes "rule-based, knowledge- based, analogy-based, statistics-based, etc." methodologies for machine translation. This indicates a diversity of ways to approach language and thus a variety of tools which could be developed to deal with the different issues posed by writing a complex text at a highly professional level.

As a basis for these tools, I would like to draw upon work in robotics by Kuniyoshi and his colleagues (Kuniyoshi, Inaba and Inoue 1994, Kuniyoshi and Nagakubo 1997, Kuniyoshi and Berthouze 1998) and their proposed model of using an "imitation process" to create a complex system that can teach itself and react in real time to real-world situations. Kuniyoshi and his colleagues have been trying to construct "a whole body humanoid robot as a research tool" in order to explore how to handle complex interactions (Kuniyoshi and Nagakubo 1997). They point out that the basic approaches to robot programming used until now do not allow for the kind of sophistication, flexibility, and robustness that might be desirable for a truly useful system (Kuniyoshi, Inaba and Inoue 1994). To overcome these issues, Kuniyoshi and his co-workers suggest developing a system which can learn by watching. Of course, they are dealing with robots and actions in the physical world, but their ideas offer much promise to the writing task under consideration here.

Let us examine some of the ideas presented. In their 1998 paper in *Neural Networks*, Kuniyoshi and Berthouze discuss their "approach for realizing a robot which can bootstrap itself towards higher complexity through embodied interaction dynamics with the environment" (Kuniyoshi and Berthouze 1998: 1259). They describe

> the dynamics of a coupled agent-environment system. The dynamics cannot be statically decomposed so that each component corresponds to a physical constituent clipped out from the agent-environment system (Beer 1997). The nervous system is no exception. If we clip it out of the agent's body and observe its activity patterns by giving a static set of test stimuli, there is no way we can relate such observation to the natural behavior of the agent. This is because the (cyclic) causal relationship (i.e. dynamics) between the stimuli and the neural activity (and its output) is broken. The tight coupling between the neural system and the body-environment system is called embodiment.

In other words, a dynamic agent cannot be separated from the environment in which it exists without destroying the relationship between the two. As an example, this is like saying, if the eye is removed from a living human body, then it can no longer be accurately observed because its relationship with its "environment" (the human body) has been destroyed.

We can apply this concept to the writing tool being discussed here, by considering the text as the coupled agent and the discourse community as the environment in which it exists. If we "clip" the text out of its discourse community to observe its "activity patterns" divorced from its natural environment, then we cannot obtain a realistic view of the text. This is because the causal relationship between the text and its community has been broken. Without the interaction, the text itself would not have any meaning and may not even come into existence. This is why only grammar patterns and technical terms alone do not suffice for the creation of a suitable professional text. The crucial interactive element is missing. In this book, the previous chapters have dealt with what seems to be texts clipped out of their environment. However, it must be remembered that the link with the environment was always present during the analyses in the form of the Moves, which express the authors' interaction with the discourse community.

Kuniyoshi and Berthouze (1998: 1259) proposed the need for "an interactionist framework unifying the abilities for robust stable behavior, quick 'on-the-fly' adaptation to novel situations, bootstrapping learning by exploring for novel dynamics and creating higher structures which support it." This means that in order to mimic the dynamic activity between the agent and its environment, constant adaptation is essential. This is not an easy task. They note that "the architecture must meet seemingly contradicting requirements; it must be adaptive to quick changes in the dynamics, but it must not drastically destroy the pre-existing behavioral capabilities, and it should coordinate and maintain the integrity of the whole set of interactions, while exploring and creating new global structures to acquire a higher level integrity" (Kuniyoshi and Berthouze 1998: 1259). Thus, the difficulty lies in balancing opposing forces to maintain equilibria while searching for ways to attain better levels which can only be done by destroying the current equilibria.

These concepts for training robots offer hints for 'training' people, a much more complex task. What I would like to argue here is that in order to tackle a complex problem, such as the teaching of professional writing, perhaps we can borrow ideas from other related fields, such as robotics. The "interactionist framework" that Kuniyoshi and Berthouze (1998: 1259) say is needed to make better robots is precisely what is needed for a viable writing-support tool: it must be adaptable to the needs of the individual writer, it must retain pre-existing linguistic expectations of the various communities; it should coordinate all these interactions while maintaining the integrity of the task underway; and yet it should also allow exploration for the construction of new structures.The final feature is particularly important in professional writing.

To achieve their goals, Kuniyoshi and Berthouze (1998: 1260) chose to use a particular class of interaction dynamics, that of "adaptive imitation of physical behavior". They state that "imitation is a unique interaction dynamics which bridges the internal observer and the external observer so that they share a common meaning, i.e. the structure of 'similarity'. [...] Although it is often easy to define clearly, imitation is quite open ended. In contrast to an arbitrary task learning, there is no need for an externally provided specific goal for each case". The target behavior can be anything. Therefore, an account for imitation ability can be very general but not too vague as it is already constrained to be 'doing a similar thing' (Kuniyoshi and Berthouze 1998: 1262). "A future version of the [robot] system might eventually break the silence after all the efforts of action recognition, plan analysis, and execution practicing, and ask a question about the instructor's ambiguous intention, in natural language" (Kuniyoshi, Inaba and Inoue 1994: 820). What is important here is that these researchers point out that the target behavior can be "anything". Recall Gosden's (1996) finding, discussed above, that his NNS informants talked of writing "with others' words". What could be useful for solving some of the problems of the NNS professional is to apply this "adaptive imitation" concept to the writing task.

As mentioned above, non-native speakers of a language, especially from cultures like Oriental ones in which the words of ancient scholars are often learned by rote, writing with others' words is 'doing a similar thing'. This idea is not entirely mistaken as what corpus linguistics is revealing is that language is used in systematic patterns. Of course, while exact imitation would be plagiarism, the other extreme of being too creative or divergent with language usage can lead to misunderstanding or rejection by the intended audience.

7.4 'A construction of the learner'

To find effective ways of serving the needs of both the NNS writer and the FGDC, some useful ideas are available from work in information technology. For example, the phrase in this section's heading comes from Cunningham et al. (1993: 45) who propose that "the textbook of the future" would not be offered in complete form to the learner but rather would be something in which the learner participates in producing. They conclude their section of a book on *Hypertext* as follows (Cunningham et al. 1993: 45):

> We believe that the most important idea presented here is the notion that the textbook of the future will be a construction of the learner, drawing upon the database and authoring, linking and customising tools provided. Instructional software will be of a different type: instead of selecting, organising and pre-

> senting content, software will provide tools that enable students to select, construct, organise and customise information from a variety of sources and representational modes.

This notion of a means of instruction that can be created together with the learner opens avenues for introducing elements that suit specific learner needs and also for including specific discourse community needs. Here the potential problem arises of an imbalance between the learner and the discourse community needs. In trying to resolve such a dilemma of how to offer freedom within frameworks, in an earlier section of the book, Cunningham et al. describe their position as follows (Cunningham et al. 1993: 29):

> Perhaps the key feature of a constructivist learning environment is that the student is in charge of constructing and testing his or her own understanding. This certainly does not imply 'anything goes', i.e. the student can study anything and any understanding developed is adequate. It also does not imply that the student is simply tossed into a sink-or-swim condition. Rather, it means that the goal of constructivist instruction is to aid the student in gaining the capability to ask relevant questions, to generate authentic contexts for the use of the knowledge to guide the interpretation of the information, to test his or her views against alternative views, and to become aware of the knowledge construction process.

In other words, there is a struggle to balance the general and the specific, the community and the individual. This notion of balancing two opposing forces appears over and over again in ESP work. In celebrating the 25th year of TESOL, the state-of-the-art essay for ESP (Johns and Dudley-Evans 1993) was subtitled "International in scope, specific in purpose". In presenting an overview of ESP in the 1990s, Dudley-Evans (1997b: 6) classifies the characteristics of the discipline into two types:

> Absolute characteristics
> - ESP is designed to meet specific needs of the learner.
> - ESP makes use of the underlying methodology and activities of the disciplines that it serves.
>
> Variable characteristics
> - ESP may be related to or designed for specific disciplines.
> - ESP may use, in specific teaching situations, a different methodology from that of General English.

There is a need for an interweaving of the community and the individual, the general and the specific, and this will be discussed below as the creation of 'negotiated' texts.

7.5 'Negotiated' texts

Here I would like to propose the idea of texts being 'negotiated' from the very start of the writing process. In the professional discourse community today, there is probably no published text that is the product of one person's sole efforts. Virtually all texts are products of negotiation. In the case of a NS, the negotiation can be at the high level of the very claims that are being proposed by the text, i.e., the conceptual level expressed by the written words, as described by Myers (1990a). But in the case of a NNS, even before negotiation can take place at this level, there are usually many language-based problems that need to be dealt with.

As an example, let us consider the process of negotiation occurring in the publishing of a research journal article. If the text were a research journal article, it would start out as a joint effort of the co-authors. Even in the rare case of a paper being written by a single author, it would most likely have had input from the writer's colleagues, sometimes in senior or junior positions. After the manuscript has reached a level suitable for a restricted form of public debut (not for publication but for consideration for publication), it can be submitted to a journal for review. At this stage, the editors and reviewers begin to make their contributions. If the work is deemed suitable for the journal and the author(s) is a non-native speaker of English, the paper will very likely be submitted to some form of English-language editing because journals demand "both correct and idiomatic English" (Information for Authors, *Analytical Biochemistry*, 1998). Novice writers should be made aware of this process of negotiation behind every published paper when they first enter the process.

Here is a specific example of 'negotiation' in operation in the construction of a text by an NNS novice to an academic discourse community. Katsumi Matsuura of Tokyo Metropolitan University (personal communication) has been trying to make students in his biological sciences department aware of the various factors involved in professional communication and offers a course which includes a description of the steps involved in the review process for journal publication. This is essential because many Japanese universities now require that the work for a Ph.D. dissertation be published (or be in the process of publication) as at least two to three research articles in recognized journals in the respcctive fields. As the leading journal is likely to be an English-language one, this means that students must exhibit a certain degree of mastery of the professional discourse in order to successfully complete their work. In trying to help his students successfully publish their papers in English-language journals, he tells them to first write as simply as possible with as clear a message as possible. Matsuura says that if the student's message is understandable, then he can find ways to help the student project it into the discourse community. His recommendations start with the preparation of data for figures and tables and the writing of very simple English sentences with one idea per sentence (Noguchi and Matsuura 2000).

This notion of negotiation is also mentioned by Johns (2000, homepage http://web.bham.ac.uk/johnstf/timeap3.htm) in describing how he works with NNS who come to him for "discussions of language points (lexical, syntactic and discoursal) that take place in one-to-one consultations when students bring us their written drafts for discussion. In many cases the revision is supported by concordance data: in other cases it employs intuition: but in all cases it depends on the joint 'negotiation' of meaning by consultant and student."

7.6 A writer's 'amanuensis'

The above discussion leads us to the need for a writing-support tool that can offer the following:

- a way to help NNS successfully select and use lexical phrases in the writing of a professional genre
- a way to aid conformation to discourse community expectations to optimize communication
- a way to retain a margin of freedom to enable creative construction of new elements

The final feature will be discussed in more detail below. Here let us consider how the first two can be realized. Rather than propose one comprehensive device, I would envision clusters of tools which can be selected, customized and organized as the writer finds necessary, as discussed above, based on the writer's awareness of what questions need to be answered and where to find these answers. In considering how to realize this, let us turn to some interesting developments in the field of machine translation.

7.6.1 Help with the lexis

As mentioned in the quote at the beginning of this chapter, tools can help us "accomplish tasks not otherwise possible" (Cunningham et al. 1993: 43). Martin Kay (1997: 3) further assures us that "A computer [...] does not dehumanize by imposing its own Orwellian stamp on the products of the human spirit and the dignity of human labor but, by taking over what is mechanical and routine, it frees human beings for what is essentially human".

As a working concept for machine translation, Kay proposes a "translator's amanuensis" – he suggests the incorporation of "simple facilities peculiar to translation" into a word processing program. He calls for a gradual taking over of functions from the periphery of translation – "little steps for little feet!" (Kay 1997: 13). Anyone who has had experience with using an automatic program for translation will realize how difficult it is – the results can elicit everything from chuckling to bafflement. However, the more controlled and domain-specific the program is, the better the result is (Melby 1997), attesting to the power of a genre-conscious approach. In fact, Lange et al. (1997) proposed machine-aided human translation (MAHT) in which they suggest the use of bilingual sentence databases, which they call "translation memories", and the use of technical sentence "skeletons" which are not complete sentences. These skeletons would be used in a *Lego*-like manner as building blocks to fill the gap between the word and the sentence level. This is similar to the Move and Step concepts presented in this book. Thus, although these proposals are related to translation, the ideas can be very effectively applied to the genre features and support measures described above in the creation of genre texts.

7.6.2 Help with discourse community expectations

In the field of computer science, Woods, who conducted research in natural language communication, presents the idea of a knowledge representation system (Woods 1988: 149):

> For many application areas, ranging from military situation assessment to medical diagnosis to a multiplicity of uses for natural language understanding and visual image interpretation, a key problem is organizing relevant knowledge in a way that makes it applicable to the problem at hand. It is one thing to have knowledge as in an encyclopedia has it – written down somewhere and perhaps indexed by subject matter. It is another thing to have assimilated knowledge so that it affects our perceptions and is fully exploited in problem solving and action. [...] A critical need, then, for many applications is a knowledge representation system that addresses the problem of invoking appropriate knowledge exactly when it is relevant to the problem. The challenge comes when the body of potential knowledge becomes large and the relevant knowledge must be efficiently extracted within time and resource limitations.

The above explanation identifies where the problem lies in professional writing by NNS. Even if they may know or learn about the necessary features of an academic genre, being able to access them in a timely fashion while creating their texts would be too much of a burden. They would be trying to consciously use lexicogrammatical features which are not as automatized as they would be for a trained native speaker of English (a novice is very likely to have similar problems). If some way could be found to relieve this burden, then more attention could be directed toward the content of the text. In order to deal with such knowledge overloading, Woods suggests (1988: 174):

> A key concept is the organization of knowledge into a taxonomic lattice of situation descriptions that can be used by a computer to perceive and characterize its current situation and identify knowledge applicable to that situation. The links between nodes in such networks can provide important processing advantages for many reasoning and perception activities. In situation recognition, the links between nodes in the network can serve as access paths to direct flows of information and processing to potentially relevant knowledge. The nodes of the lattice structure can serve as rendezvous points where markers or footprints left by controlled activation processes arising from hypothesized constituent elements of a concept can meet. This facilitates the detection of coincidences of related events, which in many cases may not be suggested in isolation, and thus enables recognition of relevant situation descriptions that would not be accessible by keyword indexing and other more typical indexing conventions.

This "taxonomic lattice of situation descriptions" is equivalent to the types of review identified in the present work with the Moves serving as the nodes. Thus, synthesis of the ideas presented in this chapter leads to an NNS writer's 'amanuensis,' which makes links possible among tools for related genres, reference frameworks, moves, and word/phrase webs.

7.7 An amanuensis for review article writing

A suggested flow chart for the process of writing a review article based on the findings of this study would begin with thinking of the title and aim of the review, planning the outline (headings) and abstract, and then deciding on the focus of the review, whether it would be mainly on the history of a field or topic, the present status of a field or project, the development of a model or theory or making the community aware of issues in a field or topic. This would be followed by employing the moves and move steps detailed in Table 47.

7.7.1 Title and statement of aim

The aim of the review should be expressed as clearly as possible in order to determine the type of review to be written. The title can be considered to be the summary of the entire paper and is one of the keys to the purpose of the review. A statement expressing the aim of the review should also be prepared to help guide the author. As a start, the writer can select one that is closest to their aim and prepare a tentative title and aim statement.

7.7.2 Section headings

Another important key to the review focus type is the section headings. The reviews do not have a set structure such as that of the research article (IMRD) and therefore section headings are valuable as a guide to their contents. The topics to be covered should be listed and tentative section headings should be prepared.

7.7.3 Abstract

By this point, the writer should have a general idea of what information the review will contain and should be able to write a tentative abstract.

7.7.4 Review focus and organization

Based on the focus of the review to be written, the next job would be to arrange the information into the Moves.

7.7.5 Move Steps

To linguistically express the Moves, the Move Steps can be used. The features and lexical phrases used for these Move Steps are described in Chapters 5 and 6 and a summary of the Step features is presented in Table 47.

7.8 Freedom for creative expression

This chapter has outlined tools that could be useful to support NNS writing for an international professional discourse community. The question then arises of how they can help expand the horizons of the science discourse community as a whole. The answer is that they can help with the more rigid linguistically based rules. Separating the different levels of writing tasks allows the writer to use convenient tools for 'lower' level tasks to allow more freedom with the 'higher' level tasks, e.g. introducing alternative rhetorical patterns in writing and through them, alternative worldviews in science. Of course, new rhetorical patterns are not likely to be accepted for the more established genres, but the review article offers an opportunity to experiment with alternative rhetorics for two reasons: (1) it is usually written on invitation by a senior researcher in the field who seems to be allowed a relative amount of freedom in the form of the final product and (2) it serves as a bridge genre between the focused discourse community and a wider audience, thus reaching a greater number of people.

This margin of freedom in the final product is evident from the analyses presented in this work. Many exceptions were noted despite the fact that all of the 25 articles were classified and published as review articles by a reputable scientific journal. First, there were differences at the physical feature level. Not all the reviews have abstracts, which would be a requisite for a research article. Despite not having a set structure, such as the Introduction, Materials and Methods, Results and Discussion sections of the research article, not all of the reviews have section headings to guide the reader. These headings, together with the title and abstract, were found to offer important clues as to the main message of the review. The length of the review articles varied greatly from about 2,000 to more than 8,000 words.

A second feature suggesting more freedom of expression for the review article was the extensive use of references to unpublished work by some of the writers. One of the specialist informants said this was probably due to the desire to offer the most up-to-date information and did not seem to think negatively about this. In fact, another specialist informant commented that the author of that review was a respected name in the field and that the review itself was one that he would use to provide pattern guidelines on how to write a review. For another review, the specialist informant was slightly critical of the author having included new experimental data but noted that the author was a leading expert in the field. Still another review devoted more than one-fourth of its Steps to describing current research being done in his laboratory. Such variation in the citation content suggests that because of the respected position of the author, even information which has not been scrutinized by the usual discourse community publication process can be accepted for a review article.

A third interesting feature was the inclusion of an oral presentation as a review. Reflecting the fact that the paper was originally prepared as a speech, it includes many direct quotes and significantly higher instances of the DiscourseSignal and Opinion Steps in comparison with the other reviews. This wide variation in content from an adapted oral presentation to the presentation of current research to a description of field history with literary quotes indicates the extent of freedom allowed.

The fourth feature was the finding that at least four types, or focus categories, of reviews could be identified for the corpus studied. Interviews with the specialist informants and analyses of the texts revealed that, while there was overlap of the moves and steps, the basic aims of the articles in the corpus could be classified into four categories: presentation of historical background of a field or a portion of it, description of the current state of a field, synthesis of field knowledge into a theory or model, and calling attention to some issue in the field. However, the instructions to authors of the journal studied and of other journals examined do not usually specify the content of the review article, very often not specifying anything except the fact that reviews are usually written on invitation. Thus, the aims of a review article can vary and may include those other than the four discussed in this work.

The audience for reviews can range from a very general one to researchers in a related field to those interested in a specific topic. This is an interesting phenomenon for a genre, which is usually considered to function within a discourse community (Bhatia (1993: 13) adapting from Swales: "a recognizable communicative event characterized by a set of communicative purpose(s) identified and mutually understood by the members of the professional or academic community in which it regularly occurs"). As the work described in this volume has shown, reviews do have a set of communicative purposes, as attested to by the specialist informant responses, but the audience does not necessarily have to be members of the immediate discourse community. In fact, sometimes the audience targeted is in a related field (TPRI trying to relate two neurodegenerative disorders) or a more general audience (such as that of the IKEY review considering ecological limits to the world's population which seems to also have policy administrators in mind). Another review which seems to be targeting an even more general audience is IAME which calls for more consumption of fruits and vegetables by Americans to improve health.

In sum, a writer's amanuensis can offer specific help for the NNS at the level of lexical phrases and organization frameworks to facilitate participation in a discourse community. This can free the writer for more creative expression in a genre such as the review article. A combination of writing support for specific surface features plus Move and Step frameworks can offer a window through which more rhetorical diversity might be introduced.

8. Towards a more open forum for science

If we examine the discourse of science we can become more aware of the ideology that is enshrined in the way scientific language construes the world.
[Halliday 1993a: 132]

Kyudou, son'i
(Finding a common path but respecting differences) [Title of an NHK (Japanese public television) program on the developing relationship between the United States and China, November 1998]

8.1 A bridge genre where 'truth comes of age'

This book has tried to show the importance of the review article in the sciences. Little work had been done on this genre (Swales 2003). The work presented here has attempted to: (1) characterize a hitherto uncharted genre which seems to offer a margin of freedom of expression and have a high discourse community impact factor, (2) describe the need for linguistic tools for NNS (nonnative speakers of English) to fully participate in their international professional discourse community, and (3) present a practical writing-support framework for the NNS scientist writer.

The deluge of information in the sciences today, expressed by several of the specialist informants consulted in this study and also discussed by Paul and Charney (1995), makes it imperative to have a means of reviewing, evaluating and synthesizing this flood of information into useful knowledge frameworks. The review article, which attempts to do this, should therefore become increasingly more important in many research fields.

The review article seems to serve as a bridge genre in several ways. First, it is a professional genre that aims to address not only the members of its own discourse community, but also those in the periphery and sometimes beyond. One of the informants, Thomas, a biochemist, said that he had been requested to write a review for a physics journal to "show physicists how physics can be applied to an important and interesting area of biology." Other informants stated that a review article was useful when entering a field in which they were not very well versed, and Lester specifically said "I know of at least two instances where a textbook followed." This suggests that the review article can also be a bridge between different levels of expertise in a community.

Still another way in which the review article can be thought of as being a bridge genre is its role in nurturing a 'fact'. Fleck (1979) and Myers (1990b) present the idea that "scientific facts" can be considered to have a "life". If this is true, perhaps it could be chronicled as follows:

STAGE	MANIFESTATION
Conception	Idea in the mind of a researcher
Birth	Written manifestation available to public scrutiny
	Notebook (lab notebooks, records, notes)
	Conference abstract
	Conference proceedings
	Research article
	Follow-up research articles
	Citation in other research articles
Coming-of-age	Acceptance as "truth" by scientific community
	Review
	Popular science literature
Middle age	Acceptance as 'truth' by general community
	Textbook
	Popular literature
Death	No longer accepted by community
or	or superceded by a new paradigm
Sainthood	Acknowledged as a 'law of nature'

The above suggests that the review article is an important stage in the life of a fact. It seems to serve as 'a coming-of-age ceremony' to make official the recognition of a fact as a portion of 'truth' and as such, allow it to be used to construct other truths (e.g. to be used in textbooks).

One example of this vital flow of ideas can be traced for research now taking place in neuroscience. The SRAM review examined in this book presents ideas related to anosognosia, or the perception of phantom limbs. This review was published in 1993, and in 1998, the author Ramachandran published a popular science book together with science writer Sandra Blakeslee (Ramachandran and Blakeslee 1998). In that book, he traces experiences that led him to undertake his current work. He chronicles the phrase "phantom limb" as first being used by the American Civil War physician Silar Weir Mitchell but notes that until only fifteen years ago, this phenomenon was denigrated in a psychiatry research journal as being "merely the result of wishful thinking", an argument which Ramachandran labels as "utter nonsense" (Ramachandran and Blakeslee 1998: 23). The conception stage for his current research is presented as follows (Ramachandran and Blakeslee 1998: 24-25):

> Like most physicians, I was intrigued by phantoms the very first time I encountered them and have been puzzled by them ever since [...] and the textbooks I consulted only deepened the mystery [...]. All these clinical experiences lay tucked away in my brain, dormant, until about six years ago, when my interest was rekindled by a scientific paper published in 1991 by Dr. Tim Pons of the National Institutes of Health, a paper that enrolled me in a whole new direction of research [...].

Note that Ramachandran states that the textbooks did not offer the information he had been seeking. After publishing a number of research reports in professional journals, some co-authored, Ramachandran wrote the review article examined in this book. The 1998 book *Phantoms in the Brain* has been acclaimed by Nobel laureate Francis Crick and was a finalist in the 1998 Los Angeles Times Book Prizes, indicating that the idea of phantom limbs has now become accepted by the scientific community in general.

8.2 The review article – a challenging genre

Because of the important functions of a review article, it is usually written by a senior (not in age or position, but experience) researcher who possesses a broad perspective of the entire field of study. It is usually written on invitation from the editor of a journal or after a request for submission has been considered and approved by the reviews editor. This is one significant difference from the research journal article which can be submitted by practically any researcher. Thus, writing a review is considered a prestigious undertaking albeit a rather challenging one.

For nonnative speakers of English, the challenge is a double one. It means producing a respected scholarly document not only from the viewpoint of content but also linguistically, in a language that is not their native tongue. Although such scientists may have published numerous journal articles and thus established their niche in the field, which led to their being asked to author a review, they may not have the linguistic skills with which to write it effectively. This is because a review differs from a research article which has evolved into an almost formulaic discourse genre. While there are 'Instructions to the Author', which are sometimes very explicit, the content and organization of the review article is usually left up to the writer. While there are books, textbooks, manuals, even computer software, to help the writer of a research article, there is little indeed on 'how to write a review'. Day (1988), in his well-known book, *How to Write and Publish a Scientific Paper*, 3rd ed., allots only five pages to it in his 178-page book (not including appendices).

8.3 The review article – an opportune genre

The dearth of information about the review article may actually be a blessing in disguise. The review article may offer an opportunity for exploring rhetorical patterns from other linguisticultures and using them to create new vistas in the world views of the scientific community. Although there are calls for this (Swales 1997, Dudley-Evans 1997a), it would be almost impossible to realize this with the research article which has a history of more than 300 years since its birth with the 'midwifery' of Henry Oldenburg of the Royal Society in England (Gross 1990: 176). The research article now has such an established format that it might be fruitless to try to change it or to introduce innovative patterns. In fact, it might not be wise to attempt this because the format in current use is the product of an enormous amount of trial-and-error and has assumed almost a 'frozen' form of efficiency. For example, there are journals which precisely state that the Abstract should contain the following labeled sections: background, aims, methods, results, and conclusions *(Gastroenterology).* Still others specify the verb tense to be used for presenting different kinds of information, and some even list certain words which should not be used in the title (*Analytical Biochemistry*, Information for Authors, July 1996):

> In creating a title for your paper, please omit words such as *novel, rapid, improved, simple, sensitive, efficient, convenient, new.* Each is felt to be redundant and actually slows down the reader who is scanning the article titles and wants to know more about the method itself. Please substitute informative words that will give readers a more complete idea of the usefulness of your paper with regard to the limitations of their equipment or sample type.

Unlike the research article which has a well established form, the review article appears to possess more fluidity. This may be because it is still in the process of being formed as a genre or that the nature of its communicative purpose allows for more diversity. Perhaps this is because it seems to serve in a bridge position between the discourse community of a specific field of research and the more general academic community. Also, examination of documents related to the writing of the review article reveals that there is much leeway for innovation, i.e., the 'rules' are not as set as they are for the research paper – the only real requirement seeming to be that the author is recognized by the discourse community as being qualified to speak for the field. This might be just what is needed to reply to the calls for more acceptance of diverse rhetorical patterns from other lingusticultures (Swales 1997, Dudley-Evans 1997a).

If science is rhetoric and rhetoric is a manner of persuasion, could there not be various ways of persuasion that might make us more aware of diverse ways of viewing this world that we live in? Halliday, in his paper analyzing Chinese and English scientific texts, concludes with the following (1993b: 132):

> What seems important to me, as a linguist and as a teacher, is that we should understand both *how* the grammar of science works and *why* it works the way it does. For this we need to examine scientific discourse, analysing the grammar in terms of its different functions; and also to trace its history, how it evolved the way it did. I asked at the beginning: would scientific Chinese have been different if modern science had evolved first in China? Obviously we can never answer this kind of question with certainty.
>
> Of course, although we can ask why things happened in the past, we cannot change them. We can, however, adopt practices that will have an effect on the future; and, as linguists, and discourse specialists, we have a responsibility to look ahead: to assist those who are learning science either in their mother tongue or through some foreign language; to advise on language planning – whether scientific language should be 'engineered' in some way; and, not least, to raise issues of the role of science, and of scientists, in our modern society. If we examine the discourse of science we can become more aware of the ideology that is enshrined in the way scientific language construes the world.

Halliday's fears are echoed by Hanappi and Egger (1997) in their assessment of information technology. They use television as an example to show how commercial messages are beguilingly packaged to make "people start to change their mental *models* of things by using a subconsciously delivered *interpretation schemata*" (1997: 195, emphasis in original). They argue that "free-choice within a set of possibilities chosen by the producer is pseudo-choice" because people become "passive receivers of preselected information" (Hanappi and Egger 1997: 194). As we enter an age governed by a new 'techno-economic paradigm', Freeman (1997: 170) points to a shift from 'centralization' to 'distributed intelligence'.

To promote diversity, the review article would be an opportune genre. It appears to be a genre with a higher tolerance for variation than its sibling genre, the research article. Therefore, if an NNS scientist were recognized as an expert in his/her field and asked to write a review, then with the aim of the review made clear using suitable expressions for the title, abstract and section headings, the author can be expected to have a margin of freedom for the text content and expression. Of course, suitable lexical phrases would need to be employed for the text to be deemed acceptable but the rhetorical features of the individual sections of the text would most likely be left to the author's discretion. It is here that, as Bhatia (1993: 13) says, adapting Swales' definition (1981, 1985, 1990a), the genre conventions can be "exploited by the expert members of the discourse community to achieve private intentions within the framework of socially recognized purpose(s)."

With the review article on the border of the domain discourse community and the wider academic discourse community and offering more expressive freedom than the research journal article, it could serve as an ideal platform where different rhetorics, different sciences might be presented to enrich our interaction with the natural world. Indeed, the review article may be an opportune genre for science, which today is being seen with much criticism and pessimism, for example, as expressed by Humphries (1998) and more seriously by Horgan (1996). John Horgan, former senior writer at *Scientific American*, who interviewed scientists from Stephen Hawking to Roger Penrose, Marvin Minsky and Noam Chomsky, wrote about "The End of Science," stating that "Belief in the eternality of progress [...] is the dominant delusion of our culture" (Horgan 1996: 268). Today there is much call for the preservation of biodiversity. Here I would like to call for a recognition of and the preservation of rhetorical diversity. Perhaps scientific discourse using alternative rhetorics might present new possibilities for the social construction of knowledge.

As rhetoric is language-based and language reflects culture, let me quote once more from Ramachandran and Blakeslee (1998: 227-228):

> In the first half of the next century, science will confront its greatest challenge in trying to answer a question that has been steeped in mysticism and metaphysics for millennia: What is the nature of the self? As someone who was born in India and raised in the Hindu tradition, I was taught that the concept of the self – the "I" within me that is aloof from the universe and engages in a lofty inspection of the world around me – is an illusion, a veil called *maya*. The search for enlightenment, I was told, consists of lifting this veil and realizing that you are really "One with the cosmos." Ironically, after extensive training in Western medicine and more than fifteen years of research on neurological patients and visual illusions, I have come to realize that there is much truth to this view [...]

In this book, I have presented a classification of types of review articles and a framework of moves and steps for analyzing review articles. Of course, this classification is only a heuristic measure to promote writing support. However, as the analyses have shown, almost all moves and steps appear in all review types and the classifications presented are supported by linguistic evidence and information from the specialist informants consulted.

I have also proposed a writer's amanuensis to help write the review article, offering suggested procedures, move frameworks, and step features. These aids, especially the step analyses, can serve as a basis for pragmatic tagging to reveal how expository texts package information. Also, as the moves composing these reviews are similar to those employed in the literature review and the discussion and conclusion sections of other academic genres, these frameworks might be useful as aids for novice writers, whether they are native or nonnative English speakers.

Finally, I hope that the ideas presented in this book can help non-native speakers of English participate more actively in their professional discourse communities, which should help enrich the discourse communities and eventually benefit our global community.

Appendix 1
Review articles examined in this book

The first three letters of the author's last name (or first author's last name in the case of multiple authors) is preceded by H for history, S for status quo, T for theory, and I for issue. These classifications are described in Chapter 3.

Abbreviated name	Reference information
HBRO	Brownstein, Michael J. (1993) A brief history of opiates, opioid peptides, and opioid receptors. *Proc. Natl. Acad. Sci. USA*, 90, 5391-5393.
HKNU	Knudson, Alfred G. (1993) Antioncogenes and human cancer. *Proc. Natl. Acad. Sci. USA*, 90, 10914-10921.
HNIC	Nicolaou, K.C., Smith, A.L. and Yue, E.W. (1993) Chemistry and biology of natural and designed enediynes. *Proc. Natl. Acad. Sci. USA*, 90, 5881-5888.
HOLS	Olson, Maynard V. (1993) The human genome project. *Proc. Natl. Acad. Sci. USA*, 90, 4338-4344.
SCLE	Clegg, Michael T. (1993) Chloroplast gene sequences and the study of plant evolution. *Proc. Natl. Acad. Sci. USA*, 90, 363-367.
SGOL	Golbeck, John H. (1993) Shared thematic elements in photochemical reaction centers. *Proc. Natl. Acad. Sci. USA*, 90, 1642-1646.
SLAI	Lai, Eseng, Clark, Kirk L., Burley, Stephen K. and Darnell, James E. Jr. (1993) Hepatocyte nuclear factor 3/fork head or "winged helix" proteins: A family of transcription factors of diverse biologic function. *Proc. Natl. Acad. Sci. USA*, 90, 10421-10423.

SLEN Lengyel, Peter (1993) Tumor-suppressor genes: News about the interferon connection. *Proc. Natl. Acad. Sci. USA*, 90, 5893-5895.

SRAM Ramachandran, V.S. (1993) Behavioral and magnetoencephalographic correlates of plasticity in the adult human brain. *Proc. Natl. Acad. Sci. USA*, 90, 10413-10420.

SSHU Shulman, R.G., Blamire, A.M., Rothman, D.L. and McCarthy, G. (1993) Nuclear magnetic resonance imaging and spectroscopy of human brain function. *Proc. Natl. Acad. Sci. USA*, 90, 3127-3133.

SVAL Vallee, Richard (1993) Molecular analysis of the microtubule motor dynein. *Proc. Natl. Acad. Sci. USA*, 90, 8769-8722.

SVAU Vaux, David L. (1993) Toward an understanding of the molecular mechanisms of physiological cell death. *Proc. Natl. Acad. Sci. USA* 90, 786-789.

SWIL Wilson, T. Michael (1993) Strategies to protect crop plants against viruses: Pathogen-derived resistance blossoms. *Proc. Natl. Acad. Sci. USA*, 90, 3134-3141.

TAND Anderson, Richard G.W. (1993) Caveolae: Where incoming and outgoing messengers meet. *Proc. Natl. Acad. Sci. USA*, 90, 10909-10913.

TBEN Bennett, Mark K. and Scheller, Richard H. (1993) The molecular machinery for secretion is conserved from yeast to neurons. *Proc. Natl. Acad. Sci. USA*, 90, 2559-2563.

TBRE Breslow, Jan L. (1993) Transgenic mouse models of lipoprotein metabolism and atherosclerosis. *Proc. Natl. Acad. Sci. USA*, 90, 8314-8313.

TGRA Grant, Verne (1993) Modes and origins of mechanical and ethological isolation in angiosperms. *Proc. Natl. Acad. Sci. USA*, 90, 3-10.

TPRI Price, Donald L., Borchelt, David R. and Sisodia, Sangram S. (1993) Alzheimer disease and the prion disorders: Amyloid β-protein and prion protein amyloidoses. *Proc. Natl. Acad. Sci. USA*, 90, 6381-6384.

TSCH Schimmel, Paul, Giegé, Richard, Moras, Dino and Yokoyama, Shigeyuki (1993) An operational RNA code for amino acids and possible relationship to genetic code. *Proc. Natl. Acad. Sci. USA*, 90, 8763-8768.

IAME Ames, Bruce N., Shigenaga, Mark K. and Hagen, Tory M. (1993) Oxidants, antioxidants, and the degenerative diseases of aging. *Proc. Natl. Acad. Sci. USA*, 90, 7915-7922.

IBEU Beutler, Ernest (1993) Gaucher disease as a paradigm of current issues regarding single gene mutations of humans. *Proc. Natl. Acad. Sci. USA*, 90, 5384-5390.

IBLE Blenis, John (1993) Signal transduction via the MAP kinases: Proceed at your own RSK. *Proc. Natl. Acad. Sci. USA*, 90, 5889-5892.

IEPS Epstein, Emanuel (1993) The anomaly of silicon in plant biology. *Proc. Natl. Acad. Sci. USA*, 90, 11-17.

IKEY Keyfitz, Nathan (1993) Are there ecological limits to population? *Proc. Natl. Acad. Sci. USA*, 90, 6895-6899.

ITEM Temin, Howard M. (1993) Retrovirus variation and reverse transcription: Abnormal strand transfers result in retrovirus genetic variation. *Proc. Natl. Acad. Sci. USA*, 90, 6900-6903.

Appendix 2
Specialist informants consulted

All were affiliated with the University of Birmingham, UK.

Informant	Text	Dept.	Position	Specialty
Al-Rubeai, Mohamed	SVAU	Chemical Engineering	Senior Fellow in Biotechnology	Cell culture
Lecane, Phil	HKNU	Cancer Research	Ph.D. student	Cancer research, tumor-suppressor genes
Lester, Richard N.	SCLE, TGRA	Biological Science	Lecturer	Botany, Systematics, esp. Biochemical taxonomy
Michell, Robert H.	HBRO	Biochemistry & CCRIS*	Professor	Biochemistry
Mitchell, Ian	SVAU	Psychology	Senior Lecturer	Cell death in the nervous system
Newbury, John	SCLE, SWIL	Biological Science	Senior Lecturer	Plant biology
Pritchard, Jeremy	IEPS	Biological Science	Lecturer	Plant physiology
Strain, Alastair	SLAI	Biochemistry	Senior Lecturer	Liver growth & differentiation
Sweet, Clive	TBRE	Biological Science	Senior Lecturer	Viral pathogenicity
Thomas, Chris M.	HOLS, TSCH	Biochemistry.	Professor	Molecular biology
Thomas, Neil	SVAL	Physics	Lecturer	Muscle biophysics
Thornhill, Rhulon Alfred	SVAL	Biological Science	Lecturer	Biophysics
Wood, Kenneth Roger	ITEM, SWIL	Biological Science	Senior Lecturer	Plant virology

* CCRIS, Centre for Clinical Research in Immunology and Signalling; The Medical School

References

American Society for Microbiology 1991. *ASM Style Manual for Journals and Books*. American Society for Microbiology.

Atkinson, Dwight 1992. The evolution of medical research writing from 1735 to 1985: The case of the Edinburgh Medical Journal. *Applied Linguistics* 13(4): 337-374.

Barnes, Gregory A. 1982. *Communication Skills for the Foreign-born Professional*. Philadelphia, PA: ISI Press.

Bazerman, Charles 1988. *Shaping Written Knowledge: The Genre and Activity of the Experimental Article in Science*. Madison, Wisconsin, USA: The University of Wisconsin Press.

Beer, Randall D. 1997. The dynamics of adaptive behavior: a research program. *Robotics and Autonomous Systems* 20: 257-289.

Belcher, Diane / Braine, George 1995. *Academic Writing in a Second Language. Essays on Research and Pedagogy*. Norwood, NJ: Ablex Publishing Corporation.

Berkenkotter, Carol / Huckin, Thomas N. 1995. *Genre Knowledge in Disciplinary Communication: Cognition/Culture/Power*. Hillsdale, NJ: Lawrence Erlbaum Associates.

Berkenkotter, Carol / Huckin, Thomas N. / Ackerman, John 1991. Social context and socially constructed texts: The initiation of a graduate student into a writing research community. In Bazerman, Charles / Paradis, James (eds.) *Textual Dynamics in the Professions*. Madison: The University of Wisconsin Press, 191-215.

Bex, Tony 1996. *Variety in Written English: Texts in Society: Societies in Text*. London: Routledge.

Bhatia, Vijay K. 1993. *Analysing Genre: Language Use in Professional Settings*. London and New York: Longman.

Biber, Douglas 1988. *Variation across Speech and Writing*. Cambridge, UK: Cambridge University Press.

Biber, Douglas / Johansson, Stig / Leech, Geoffrey / Conrad, Susan / Finegan, Edward 1999. *Longman Grammar of Spoken and Written English*. Harlow, Essex: Pearson Education Limited.

Birch, Thomas 1968. The history of the Royal Society for improving of knowledge from its first rise. In Hall, A. Rupert/Hall, Marie Boas (eds.) *A Facsimile of the London Edition of 1757-57*, 4 vols., New York: Johnson Reprint.

Bizzell, Patricia 1992. *Academic Discourse and Critical Consciousness*. Pittsburgh and London: University of Pittsburgh Press.

Bley-Vroman, Robert/Selinker, Larry 1984. Research design in rhetorical/grammatical studies: a proposed optimal research strategy. *English for Specific Purposes*. Oregon State University, 82-83: 1-4 and 86: 1-6.

Bloch, Joel/Chi, Lan 1995. A comparison of the use of citations in Chinese and English academic discourse. In Belcher, Diane/Braine, George (eds.) *Academic Writing in a Second Language: Essays on Research and Pedagogy*. Norwood, NJ: Ablex, 231-275.

Brufee, Ken A. 1986. Social construction, language, and the authority of knowledge: A bibliographical essay. *College English*, 48(8), 773-790.

Burns, T. 1969. Models, images and myths in Marquis, E.G./Gruber, W.H. (eds.) *Factors in the Transfer of Technology*. Cambridge, Massachusetts: M.I.T. Press.

Campbell, Cherry 1990. Writing with others' words: Using background reading texts in academic compositions. In Kroll, Barbara (ed.) *Second Language Writing: Research Insights for the Classroom*. Cambridge: Cambridge University Press, 211-230.

Carson, J.G. 1992. Becoming biliterate: First language influences. *Journal of Second Language Writing* 1(1): 37-60.

Channell, Joanna 1994. *Vague Language*. Oxford: Oxford University Press.

Chen, Kuang-Jung 1997. *English versus Chinese: World views and writing styles. TESOL Newsletter* 7, 2: 13, 16.

Christie, Frances 1994. Developing and educational linguistics for English language teaching: A systemic functional linguistic perspective. *Functions of Language*, 1(1), 95-127.

Collins, Harry M. 1982. Tacit knowledge and scientific networks. *In* Barnes, Barry/Edge, David (eds.) *Science in Context: Readings in the Sociology of Science*. Milton Keynes, UK: The Open University Press, 44-64.

Connor, Ulla 1987. Argumentative patterns in student essays: Cross-cultural differences. In Connor, Ulla/Kaplan, Robert B. (eds.) *Writing across Languages: Analysis of L2 Text*. Reading, MA: Addison-Wesley, 57-71.

Connor, Ulla 1996. *Contrastive Rhetoric: Cross-cultural Aspects of Second-language Writing*. Cambridge: Cambridge University Press

Connor, Ulla/Lauer, Janice 1988. Cross-cultural variation in persuasive student writing. In Purves, Alan C. (ed.) *Writing across Languages and Cultures: Issues in Contrastive Rhetoric*. Newbury Park: Sage Publications, 138-159.

Council of Biology Editors Style Manual Committee 1983. *CBE Style Manual: A Guide for Authors, Editors and Publishers in the Biological Sciences*. Bethesda, MD: Council of Biology Editors, Inc.

Cunningham, Donald J./Duffy, Thomas M./Knuth, Randy A. 1993. In McKnight, Cliff/Dillon, Andrew/Richardson, John (eds.) 1993. *Hypertext: A Psychological Perspective*. New York: Ellis Horwood, 19-49.

Davies, Florence 1988. Reading between the lines: Thematic choice as a device for presenting writer viewpoint in academic discourse. *The ESP, Sao Paulo*, 9 (1/2), 173-200.

Day, Robert A. 1988. *How to Write and Publish a Scientific Pape*r (3rd ed.). Phoenix and New York: Oryx Press.

Dennett, Joann T. 1988. Not to say is better than to say: How rhetorical structure reflects cultural context in Japanese-English technical writing. *IEEE Transactions on Professional Communication*, 31(3), 116-119.

Devitt, Amy J. 1996. Genre, genres, and the teaching of genre. *College, Composition and Communication*, 47, 605-616

Douglas, Mary 1966. Purity and danger: an analysis of concepts of pollution and taboo. London: Routledge & Kegan Paul.

Dudley-Evans, Tony 1986. Genre analysis: An investigation of the introduction and discussion sections of MSc dissertations. In Coulthard, Malcolm (ed.) *Talking about Text*. Birmingham, UK: English Language Research, University of Birmingham, 128-145.

Dudley-Evans, Anthony 1995. Common-core and specific approaches to the teaching of academic writing. In Belcher, Diane/Braine, George (eds.) *Academic Writing in a Second Language: Essays on Research and Pedagogy*. Norwood, NJ: Ablex, 293-312.

Dudley-Evans, Tony 1997a. Genre: how far can we, should we go? *World Englishes*, 16, 3: 351-358.

Dudley-Evans, Tony 1997b. An overview of ESP in the 1990s. In Orr, Thomas (ed.) *Proceedings 1997: The Japan Conference on English for Specific Purposes*. University of Aizu, Fukushima, Japan, 5-11.

Dudley-Evans, Tony/St. John, Maggie Jo 1998. *Developments in English for Specific Purposes*. Cambridge: Cambridge University Press.

Edge, Julian/Richards, K. (1998). May I see your warrant, please?: Justifying outcomes in qualitative research. *Applied Linguistics* 19(3): 334-356.

Eggington, W.G. 1987. Written academic discourse in Korean: Implications for effective communication. In Connor, Ulla/Kaplan, Robert B. (eds.) *Writing across Languages: Analysis of L2 Text*. Reading, MA: Addison-Wesley, 153-168.

Elbow, Peter 1991. Reflections on academic discourse. *College English*, 53(2), 135-155.

Fahnestock, Jeanne 1986. Accommodating science: The rhetorical life of scientific facts. *Written Communication*, 3: 3, 275-296.

Fanning, P. 1977. Review of Brasnett (C. Brasnett (1969) English for engineers, Methuen). *ESPMENA Bulletin* 9.

Fleck, Ludwik 1979. *Genesis and Development of a Scientific Fact*. Trenn, T.J./Merton, R.K. (eds.) Chicago and London: The University of Chicago Press.

Freeman, Chris 1997. The political economy of the long wave. In Tylecote, Andrew/Van der Straaten, Jan (eds.) *Environment, Technology and Economic Growth: The Challenge to Sustainable Development*. Cheltenham: Edward Elgar, 159-183.

Freeman, Walter J. 1991. The physiology of perception. *Scientific American*, February, 34-41.

Fujimoto, W. 1996. Linguisticulture today. In *Proceedings of the International Symposium on "Linguisticulture: Where do we go from*

here?" December 16, 1996, Graduate School of Language and Culture, Osaka University. 91-93.
Geertz, Clifford 1988. *Words and Lives: The Anthropologist as Author*. Palo Alto, CA: Stanford University Press.
Gibbs, W. Wayt 1995. Lost science in the Third World. *Scientific American*, 273, 2: 92-99.
Gibbs, W. Wayt 1997. Taking computers to task. *Scientific American*, 277, 1: 64-71.
Gilbert, G. Nigel / Mulkay, Michael 1982. Warranting scientific belief. *Social Studies of Science*, 12, 383-408.
Gosden, Hugh 1992a. Discourse functions of marked theme in scientific research articles. *English for Specific Purposes*, 11, 207-224.
Gosden, Hugh 1992b. Research writing and NNSs: From the editors. *Journal of Second Language Writing*, 1(2), 123-139.
Gosden, Hugh 1993. Discourse functions of subject in scientific research articles. *Applied Linguistics*, 14(1), 56-75.
Gosden, Hugh 1996. Verbal reports of Japanese novices' research writing practices in English. *Journal of Second Language Writing*, 5(2), 109-128.
Gross, Alan G. 1990 *The Rhetoric of Science*. Cambridge, MA: Harvard University Press.
Gunawardena, Charlotte N. 1989. The present perfect in the rhetorical divisions of biology and biochemistry journal articles. *English for Specific Purposes*, 8, 265-273.
Hagstrom, Warren O. 1982. Gift giving as an organizing principle in science. In Barnes, Barry / Edge, David (eds.), *Science in Context: Readings in the Sociology of Science*. Milton Keynes, UK: The Open University Press, 21-34.
Halliday, Michael A.K. 1993a. *Language in a Changing World*. Applied Linguistics Association of Australia.
Halliday, Michael A.K. 1993b. The analysis of scientific texts in English and Chinese. In Halliday, Michael A.K. / Martin, Jim R. (eds.) *Writing Science: Literacy and Discursive Power*. London, Washington, D.C.: The Falmer Press, 124-132.
Halliday, Michael A.K. / Hasan, Ruqaiya 1985. *Language, Context, and Text: Aspects of Language in a Social-semiotic Perspective*. Oxford University Press.

Halliday, Michael A.K./Martin, Jim R. 1993. *Writing science: Literacy and Discursive Power*. London, Washington, D.C.: The Falmer Press.

Halliday, Michael A.K./MacIntosh, Angus/Strevens, Peter 1964. *The Linguistic Sciences and Language Teaching*. London: Longman.

Hanappi, Gerhard/Egger, E. 1997. Information age – deformation age – reformation age: An assessment of the information technology Kondratieff. In Tylecote, Andrew/Van der Straaten, Jan (eds.) *Environment, Technology and Economic Growth: The Challenge to Sustainable Development*. Cheltenham: Edward Elgar, 184-205.

Harwood, John 1986. Review: Ludwik Fleck and the sociology of knowledge. *Social Studies of Science*, 16, 173-187.

Hesse, Mary 1980. *Revolutions and Reconstructions in the Philosophy of Science*. Brighton, Sussex: The Harvester Press.

Hinds, John 1983. Topic continuity in Japanese. In Givon, Talmy (ed.) *Topic Continuity in Discourse: A Quantitative Cross-language Study*. Amsterdam and Philadelphia: John Benjamins Publishing Co., 47-93.

Hinds, John 1986. Contrastive rhetoric: Japanese and English. *Text* 3, 2: 183-195.

Hinds, John 1987. Reader versus writer responsibility: a new typology. In Connor, Ulla/Kaplan, Robert B. (eds.) Writing Across Languages: Analysis of L2 Text. Reading, MA: Addison-Wesley, 141-152.

Hoey, Michael 1979. *Signalling in Discourse*. English Language Research Monograph, 6, Birmingham: University of Birmingham.

Hoey, Michael 1983, 1991. *On the Surface of Discourse*. Nottingham, UK: Department of English Studies, University of Nottingham.

Holquist, Michael (ed.) 1981. *The Dialogic Imagination: Four Essays by M.M. Bakhtin* (Translation of *Voprosy literatury i estetiki*). Austin, TX: University of Texas Press.

Hopkins, Andy/Dudley-Evans, Tony 1988. A genre-based investigation of the discussion sections in articles and dissertations. *English for Specific Purposes*, 7, 113-121.

Horgan, John 1996. *The End of Science: Facing the Limits of Knowledge in the Twilight of the Scientific Age*. London: Abacus.

Hozayen, Ghada 1994. *A Study of the Discussion Section of the Medical Research Article: A Genre Based Approach to Text Analysis*. Ph.D. dissertation, University of Birmingham.
Huckin, Thomas/Olsen, Leslie 1991. *Technical Writing and Professional Communication for Nonnative Speakers of English* (2nd ed). New York: McGraw-Hill.
Humphries, Stella 1998. ESP news: An inside look at scientific language. *TESOL Matters*, February/March 1998, 9.
Hunston, Susan 1989. *Evaluation in Experimental Research Articles* Vol. I and II. Ph.D. dissertation, Department of English, Faculty of Arts, University of Birmingham.
Hunston, Susan. 1993. Evaluation and ideology in scientific writing. In Ghadessy, Mohsen (ed.) *Register Analysis: Theory and Practice*. London: Routledge, 57-73.
Institute for Scientific Information. *Journal Citation Reports* 1993.
Ishiwata, T. 1971. Gendai no goi [Contemporary vocabulary]. In Sakakura, A. (ed.), *Koza kokugo shi: goi shi* [Seminar, history of Japanese: history of vocabulary], 347-411. Tokyo: Taishukan Shoten.
Johns, Ann M. 1993. Written argumentation for real audiences: Suggestions for teacher research and classroom practice. *TESOL Quarterly*, 27(1), 75-90.
Johns, Ann M. 1997. *Text, Role, and Context*. Cambridge: Cambridge University Press.
Johns, Ann M./Dudley-Evans, Tony 1993. English for specific purposes: International in scope, specific in purpose. In Silberstein, Sandra (ed.) *State of the art TESOL essays: Celebrating 25 years of the discipline*. Alexandria, VA: TESOL, 115-132.
Johns, Tim 2000. homepage
http://web.bham.ac.uk/johnstf/timeap3.htm
Johns, Tim F./Dudley-Evans, Anthony 1980. An experiment in team-teaching of overseas postgraduate students of transportation and plant biology. In *Team teaching in ESP. ELT Documents* 106.
Kanaoka, Masao 1999. A technical writing course aimed at nurturing critical thinking skills. *The Language Teacher* 23(9): 16-19 (+37).
Kaplan, Robert 1966. Cultural thought patterns in intercultural education. *Language Learning*, 16, 1-20.

Kaplan, Robert 1987. Cultural thought patterns revisited. In Connor, Ulla / Kaplan, Robert B. (eds.) *Writing across Languages: Analysis of L2 Text*. Reading, MA: Addison-Wesley, 9-21.
Kaplan, Robert B. 1988. Contrastive rhetoric and second language learning: Notes toward a theory of contrastive rhetoric. In Purves, Alan C. (ed.) *Writing across Languages and Cultures: Issues in Contrastive Rhetoric*. Newbury Park: Sage Publications, 275-304.
Kauffman, Stuart A. 1991. Antichaos and adaptation. *Scientific American*, August, 64 -70.
Kay, Martin (1997) The proper place of men and machines in language translation. *Machine Translation*, 12, 3-23.
Killingsworth, M. Jimmie / Gilbertson, Michael K. 1992. *Signs, Genres, and Communities in Technical Communication*. Amityville NY: Baywood Publishing Company.
Kinoshita, Koreo 1981. *Rikakei no sakubun gijutsu* (Writing techniques for science, in Japanese). Tokyo: Chuokoronsha.
Kiritani, Elizabeth 1997. My Nippon: Weighing the need to know. *The Daily Yomiuri*, September 27, 1997, 8.
Knorr Cetina, Karin D. 1981. *The Manufacture of Knowledge*. Oxford: Pergamon Press.
Knorr Cetina, Karin 1995. Metaphors in the scientific laboratory: Why are they there and what do they do? In Radman, Zdravko (ed.) *From a Metaphorical Point of View: A Multidisciplinary Approach to the Cognitive Content of Metaphor*. Berlin, New York: Walter de Gruyter, 329-349.
Kobayashi, Hiroe 1984. *Rhetorical Patterns in English and Japanese*. Unpublished Ph. D. dissertation, Columbia University, New York, U.S.A.
Kress, Gunther / Threadgold, Terry 1988. Towards a social theory of genre. Southern Review, 21/3, 215-43.
Kubota, Ryuko 1997. A reevaluation of the uniqueness of Japanese written discourse: implications for contrastive rhetoric. *Written Communication*, 14, 4: 460-480.
Kuhn, Thomas S. 1962. *The Structure of Scientific Revolutions*. Chicago: University of Chicago Press.
Kuniyoshi, Yasuo / Inaba, Masayuki / Inoue, Hirochika 1994. Learning by watching: extracting reusable task knowledge from visual

observation of human performance. *IEEE Transactions on Robotics and Automation*, 10, 6: 799-822.

Kuniyoshi, Yasuo / Nagakubo, Akihiko 1997. Humanoid interaction research: Exploring meaningful order in complex interactions. *Proceedings of the International Conference on Complex Systems* (NH, Sept. 21-26, 1997).

Kuniyoshi, Yasuo / Berthouze, Luc 1998. Neural learning of embodied interaction dynamics. *Neural Networks* 11: 1259-1276.

Lackstrom, John E. / Selinker, Larry / Trimble, Louis P. 1972. Grammar and technical English. English Teaching Forum. X, 5. Reprinted in Swales, John (1985) *Episodes in ESP*. Oxford: Pergamon, 58-68.

Lange, Jan-Matti / Gaussier, Eric / Daille, Béatrice 1997. Bricks and skeletons: some ideas for the near future of MAHT. *Machine Translation*, 12, 39-51.

Latour, Bruno / Woolgar, Steve 1982. The cycle of credibility. In Barnes, Barry / Edge, David (eds.), *Science in Context: Readings in the Sociology of Science*. Milton Keynes, UK: The Open University Press, 35-43.

Latour, Bruno / Woolgar, Steve 1986. *Laboratory Life: The Construction of Scientific Facts*. Princeton, New Jersey: Princeton University Press.

Law, John / Williams, R. J. 1982. Putting facts together: A study of scientific persuasion. *Social Studies of Science*, 12, 535-558.

Leckie-Tarry, Helen 1995. David Birch (ed.) *Language and Context: A Functional Linguistic Theory of Register*. London and New York: Pinter.

Lenat, Douglas B. 1995. Artificial intelligence: a crucial storehouse of commonsense knowledge is now taking shape. *Scientific American*, 273, 3: 62-64.

Leydesdorff, Loet 1991. In search of epistemic networks. *Social Studies of Science*, 21: 75-110.

Lynch, M. E. 1982. Technical work and critical inquiry: Investigations in a scientific laboratory. *Social Studies of Science*, 12, 499-533.

Malmkjaer, Kirsten (ed.) 1991. *The Linguistics Encyclopedia*. London and New York: Routledge.

Martin, Jim R. 1984. Language, register and genre. In Christie, Frances (ed.) *Language Studies: Children's Writing: A Reader*. Geelong: Deakin University Press, 21-30.

Mauranen, Anna 1993a. *Cultural Differences in Academic Rhetoric: A Textlinguistic Study.* Frankfurt am Main: Peter Lang.

Mauranen, Anna 1993b. Theme and prospection in written discourse. In Baker, Mona/Francis, Gill/Tognini-Bonelli, Elena (eds.) *Text and Technology: In Honour of John Sinclair*. Philadelphia and Amsterdam: John Benjamins Publishing Company, 95-114.

Maynard, Senko K. 1996. Presentation of one's view in Japanese newspaper columns: commentary strategies and sequencing. *Text* 16, 2: 391-421.

McCawley, James D. 1993. *Everything that Linguists have Always Wanted to Know about Logic – But were Ashamed to Ask* (2nd ed.). Chicago, The University of Chicago Press.

McKay, Sandra L. 1993a. Examining L2 composition ideology: A look at literacy. *Journal of Second Language Writing* 2(1): 65-81.

McKay, Sandra L. 1993b. *Agendas for Second Language Literacy.* New York: Cambridge University Press.

McKenna, Bernard 1997. How engineers write: An empirical study of engineering report writing. *Applied Linguistics*, Vol. 18, No. 2, 189-211.

Melby, Alan 1997. Some notes on the proper place of men and machines in language translation. *Machine Translation*, 12, 29-34.

Merton, Robert K. 1968. The Matthew effect in science, *Science*, 159 (3810): 56-63.

Michaelson, Herbert B. 1982. *How to Write and Publish Engineering Papers and Repo*rts. Philadelphia, PA: ISI Press.

Miller, Carolyn R. 1984. Genre as social action. *Quarterly Journal of Speech*, 70, 151-167.

Miller, Carolyn R. 1992. Rhetoric community: the cultural basis of genre. Paper presented at the Re-thinking Genre Seminar, Carlton University, Ottawa.

Muramatsu, Mieko 1997. An interview with Ulla Connor. *The Language Teacher*, 21, 4: 7-8, 39.

Myers, Greg 1985. Texts as knowledge claims: The social construction of two biology articles. *Social Studies of Science*, 15, 593-630.

Myers, Greg 1986. Review: Writing research and the sociology of scientific knowledge: A review of three new books. *College English*, 48(6), 595-610.
Myers, Greg 1988. Reviews: Writing about *Writing about Scientific Writing; Knowledge and Reflexivity; Discourse and Social Psychology; Science: The Very Idea*; and *Science in Action. College Composition and Communication*, 39, 465-474.
Myers, Greg 1989. The pragmatics of politeness in scientific articles. *Applied Linguistics*, 10(1), 1-35.
Myers, Greg 1990a. *Writing biology: Texts in the Social Construction of Scientific Knowledge*. Madison, Wisconsin: The University of Wisconsin Press.
Myers, Greg 1990b. Making a discovery: Narrative of split genes. In Nash, Christopher (ed.) *Narrative in Culture: The Uses of Storytelling in the Sciences, Philosophy, and Literature*. London and New York: Routledge, 102-126.
Myers, Greg 1991. Stories and styles in two molecular biology review articles. In Bazerman, Charles/Paradis, James (eds.) *Textual Dynamics in the Professions*. Madison: The University of Wisconsin Press.
Nagano, M. 1983. Danwa ni okeru jojutsu no koozoo [The structure of predicate in discourse]. In *Danwa no Kenkyuu to Kyooiku* [Study of Discourse and Education], Minami, F. and Tanaka, N. (eds.) Vol. 1. Tokyo: Ookurashoo Insatsukyoku, 63-89.
Nakamura, Hajime 1964. *Ways of Thinking of Eastern Peoples: India – China – Tibet – Japan*. Honolulu, Hawaii: The University of Hawaii Press.
Nakane, Chie 1970. *Japanese Society*. Berkeley CA: University of California Press.
Nakayama, Shigeru 1984. *Academic and Scientific Traditions in China, Japan, and the West*. (translated by J. Dusenbury) Tokyo: University of Tokyo Press.
Naoumi, Evelyn J. 1997. *A Cross-cultural Analysis of Editorials*. Unpublished MA dissertation. University of Birmingham.
Nattinger, James R./DeCarrico, Jeanette S. 1992. *Lexical Phrases and Language Teaching*. Oxford: Oxford University Press.

Noguchi, Judy/Matsuura, Katsumi 2000. *Judy-sensei no eigo kagaku ronbun no kakikata*. (How to write research articles in English, in Japanese) Tokyo: Kodansha.
Okamura, Akiko/Shaw, Philip 2000. Lexical phrases, culture, and subculture in transactional letter writing. *English for Specific Purposes*, 19(2000), 1-15.
Paltridge, Brian 1996. Genre, text type, and the language learning classroom. *ELT Journal* Vol. 50/3 (July 1996) 237-243.
Pandalai, S.G. 2000. February 7, 2000. http://www.transworldresearch.com/
Parker, Sybil P., ed. 1994. *McGraw-Hill Dictionary of Scientific and Technical Terms*. New York: McGraw-Hill.
Paul, Danette/Charney, Davida 1995. Introducing chaos (theory) into science and engineering: Effects of rhetorical strategies on scientific readers. *Written Communication* 12(4): 396-438.
Peirce, Charles S. 1902. Vagueness. In M. Baldwin (ed.) *Dictionary of Philosophy and Psychology II*. London: Macmillan.
Pellegrino, James 1995. Technology in support of critical thinking. *Teaching of Psychology*, 22(1), 9-10.
Porter, James E. 1992. *Audience and Rhetoric: An Archaeological Composition of the Discourse Community*. Englewood Cliff, NJ: Prentice-Hall.
Price, Derek de Solla. 1963. *Little Science, Big Science*. New York: Columbia University Press. In Collins, Harry M. (1982) *Tacit Knowledge and Scientific Net*works.
Purves, Alan C. 1990. *The Scribal Society: An Essay on Literacy and Schooling in the Information Age*. New York: Longman.
Ramachandran, Vilanyanur S./Blakeslee, Sandra 1998. *Phantoms in the Brain: Probing the Mysteries of the Human Mind*. New York: William Morrow.
Reid, I. (ed.) 1987. *The Place of Genre in Learning*. Geelong, Victoria: Deakin University Press.
Robinson, Pauline C. 1980. *ESP (English for Specific Purposes): The Present Position*. Oxford: Pergamon Press.
Ruch, W.V. 1984. *Corporate Communications: A Comparison of Japanese and American Practices*, Westport, CT Quorum Books.
Scott, Mike. *Wordsmith Version 2.0*. Oxford: Oxford University Press.

Shibata, Takeshi 1992. *Joohooka jidai no bunshoo* [Discourse of the information age]. Nihongogaku 11 (April): 8-11.

Sinclair, John / Coulthard, Malcolm 1992. Towards an analysis of discourse. In Coulthard, Malcolm (ed.) *Advances in Spoken discourse Analysis*. London and New York: Routledge, 1-34.

Sinclair, John M. / Coulthard, R. Malcolm 1975. *Towards an Analysis of Discourse: The English used by Teachers and Pupils*. London: Oxford University Press.

Stonier, T. 1984. The knowledge industry. In Forsyth, R. (ed.) *Expert Systems: Principles and Case Studies*. London and New York: Chapman and Hall, 211-226.

Swales, John 1981. *Aspects of Article Introductions*. Aston ESP Research Reports No. 1, The Language Studies Unit, The University of Aston in Birmingham.

Swales, John 1985. *Episodes in ESP*. Oxford: Pergamon Institute of English.

Swales, John M. 1990a. Discourse analysis in professional contexts. *Annual Review of Applied Linguistics*, 11, 103-114.

Swales, John M. 1990b. Genre *Analysis: English in Academic and Research Settings*. Cambridge, UK: Cambridge University Press.

Swales, John 1997. English as Tyrannosaurus rex. *World Englishes*, 16, 3: 373-382.

Swales, John 1998. *Other Floors, Other Voices: A Textography of a Small University Building*. Mahwah, NJ: Lawrence Erlbaum Associates, Publishers.

Swales, John M. 2003. "That master narrative of our time": The research article revisited. The 14th European Symposium on Language for Special Purposes: Communication, culture, knowledge. August 18-22, 2003, University of Surrey, Guildford, U.K.

Swales, John / Feak, Christine A. 1994. *Academic Writing for Graduate Students*. Ann Arbor: The University of Michigan Press.

Tadros, Angele 1981. *Linguistic Prediction in Economics Texts*. Unpublished PhD dissertation, University of Birmingham.

Teubert, Wolfgang 1996. Editorial. *International Journal of Corpus Linguistics*, 1(1), iii-x.

Ure, Jean / Ellis, Jeffrey 1977. Register in descriptive linguistics and linguistic sociology. In Uribe-Villas, Oscar 1977. Issues in Sociolinguistics. The Hague: Mouton.

Ventola, Eija 1987. *The Structure of Social Interaction. A Systemic Approach to the Semiotics of Service Encounters.* London: Francis Pinter.

Vygotsky, Lev S. 1987. *Mind and Society: The Development of Higher Psychological Processes.* Cambridge, MA: Harvard University Press.

Weissberg, Robert/Buker, Suzanne 1990. *Writing Up Research: Experimental Research Report Writing for Students of English.* Englewood Cliffs, NJ: Prentice-Hall.

Winter, Eugene 1977. A clause relational approach to English texts: A study of some predictive lexical elements in written discourse. *Instructional Science* 6/1.

Wood, Alistair 2001. International scientific English: The language of research scientists around the world. In Flowerdew, John/Peacock, Matthew (eds.) *Research Perspectives on English for Academic Purposes*, 71-83.

Woods, William A. 1988. Knowledge representation. In Bartee, Thomas C. (ed.) *Expert Systems and Artificial Intelligence: Applications and Management.* Indianapolis, IN: Howard W. Sams & Co., a division of Macmillan, Inc., 149-177.

Woolgar, Steve W. 1976. Writing an intellectual history of scientific development: The use of discovery accounts. *Social Studies of Science*, 6, 395-422.

Yates, JoAnne/Orlikowski, Wanda 1992. Genres of organizational communication: A structurational approach. *Academy of Management Review*, 17: 299-326

Yomiuri Shimbun 1999. More Japanese papers published in Nature (May 20, 1999, p. 3) Tokyo: *The Yomiuri Shimbun*.

Yunick, Stanley 1997. Genres, registers and sociolinguistics. *World Englishes*, 16 (3): 321-336.

Zappen, James P. 1991. Scientific rhetoric in the nineteenth and early twentieth centuries: Herbert Spencer, Thomas H. Huxley, and John Dewey. In Bazerman, Charles/Paradis, James (eds.) *Textual Dynamics of the Professions: Historical and Contemporary Studies of Writing in Professional Communities*. Madison, Wisconsin: The University of Wisconsin Press, 256-278.

Index

academic discourse 79
academic discourse community 97
academic prose 52
Action 41, 43, 44, 47, 48, 74, 86
adaptive imitation 231
Aim Step(s) 143, 146, 147, 150, 158, 164, 165, 166, 172, 200
alternative rhetorics 60, 62, 249
amanuensis 235, 237, 241, 250
ancient China 23
ancient Greece 23
AssessFinding Step 190, 191, 204
Bakhtin, M.M. 32
Bazerman, Charles 35, 39, 75
Belcher, Diane and Braine, George 52
Berkenkotter, Carol and Huckin, Thomas N. 30, 33, 34, 36, 39, 41, 44, 47, 48, 81, 86
Bhatia, Vijay K. 37, 44, 241, 249
Biber, Douglas 42, 51
bibliometric evaluation 99
bridge genre(s) 40, 41, 42, 97, 243, 244
chaos 24, 34, 224
"cherished" texts 31, 32
cited half-life 92
claim 20
Collins, Harry M. 19, 20, 81
components of review articles 221
concordance lines 165
Confucian rhetoric 72
Connor, Ulla 62
construction of knowledge 89, 92, 249
contrastive rhetoric 63, 79
corpus 97, 99
cross-cultural issues 58
cross-cultural rhetoric 78
CurrentResearch Step 144, 155, 182, 183, 184, 202
Define (Def) Step(s) 157, 174, 175, 201
Development (Dev) Step 205
development of genres 83
disciplinary community 26
Discourse 100
discourse community(ies) 14, 20, 25, 26, 27, 28, 29, 31, 34, 41, 44, 46, 60, 61, 62, 75, 77, 81, 92, 93, 94, 96, 119, 235, 236, 241, 248, 249, 250
    discourse community ownership of genres 30
    discourse community 'jurisdiction' 48
    multiple discourse communities 28
Discourse Organizer (DiscourseOrg) Step(s) 128, 136, 138, 142, 147, 161, 150, 151, 163, 164, 167, 171, 200, 201, 221
    DiscourseSignal Step 168, 169, 170
discovery 20, 21
Dmacro (Macro discourse entity) function 150, 166, 167
Dudley-Evans, Tony 76, 77, 78, 87, 125, 136
Ela (Elaborate Step) 185, 203
English for science and technology 37
English for specific purposes (ESP) 37, 76
English-speaking academic community (ESAC) 52, 54, 55, 56, 224
English-speaking community (ESC) 51, 54, 55, 224
Evaluating 153
Evaluating and Synthesizing Steps 140, 143, 148
evolving genre 81

exploiting genres 44
explosion of information 90, 91
Focused discourse community(ies) 29, 30, 42
Focused global discourse community (FGDC) 28, 29, 32, 51, 53, 54, 55, 59, 86, 93, 224
fora 27
Form 41, 43, 47, 48, 74, 86
forum 27
framework of moves and steps 250
freedom for creative expression 239
frequency list 170
frozen 62, 247
frozen genre 60, 61
Gap Step(s) 149, 192, 204
gatekeeping 44, 48
generic structure 38
genre(s) 14, 29, 30, 36, 37, 38, 39, 40, 41, 42, 46, 47, 48, 50, 51, 62, 74, 75, 76, 77, 78, 82, 93, 94, 241, 243
    genre action 32
    genre analysis 225
    genre evolution 33, 76
    genre features 32
    genre ownership 35
    opportune genre 94, 247, 248
gestalt(s) 24, 25
Global discourse communities 27
Gosden, Hugh 61, 129, 130, 132, 225
grammatical subject (GS) 128, 146
Greek rhetoric 73
Gross, Alan 19, 21, 23, 45, 61, 75
GS (grammatical subject) 132, 147, 148
Halliday, Michael A.K. 25, 36, 37, 38
Hinds, John 62, 64, 65
History review(s) 100, 101, 105, 107, 109, 113, 114, 119, 120, 150, 151, 153, 154, 175, 207, 208, 210
History Step(s) 154, 175, 176, 202
HistoryHypo Step(s) 188, 203
HOEmp (Empty hypothesized/ objectivized theme) 147, 149
Humphries, Stella 21
Hypothesize (Hypo) Step(s) 145, 155, 156, 158, 196, 204
impact factor 92
informant interviews 115, 118
information explosion 95
interviews 109
invisible college 19, 26
irrationality 67, 69
Issue review(s) 100, 102, 104, 106, 109, 108, 119, 120, 158, 218, 220
Johns, Ann M. 58
Journal Citation Reports (JCR) 92, 93, 98, 99
journal editors 61
journal instructions 6, 82, 92, 207
kairos 33
Kaplan, Robert B. 62, 63, 78
KeyWords 164
    KeyWord(s) list 161, 162
    Keywords search 163
ki-shoo-ten-ketsu 64, 65, 66
Killingsworth, M. Jimmie and Gilbertson, Michael K. 27, 29, 46
Knorr Cetina, Karin D. 19, 33, 35, 36
knowledge 75
Kobayashi, Hiroe 66
Kubota, Ryuko 66
Kuhn's theory of scientific revolutions 19
Latour, Bruno and Woolgar, Steve 44
lexical patterns 227
lexical phrases 226, 235, 241
Leydesdorff, Loet 50
linguisticulture(s) 67, 69, 75, 76, 248
Literature (Lit) Step(s) 144, 152, 203
Literature Step 149, 154, 156, 185, 187
LiteratureHist Step 149, 152, 187, 188
LitH (LiteratureHist) Step 147, 203
logic 68
Martin, Jim R. 25
'Matthew effect' 79
Mauranen, Anna 14, 28, 29, 30, 39, 58, 63, 75, 76, 77, 78

Merton, Robert K. 79
Miller, Carolyn 33, 35, 39, 40, 42, 44
Modal verb 145, 146
Move(s) 125, 126, 128, 221
 Move and Step frameworks 241
 Move Presenting Steps 144
Move Step(s) 100, 128, 161, 200, 221
 evaluating 137, 188, 189, 192
 presenting 137, 174, 175, 177, 179, 182, 185, 187
 synthesizing 137, 195, 198, 199
Myers, Greg 14, 20, 22, 24, 61
Nakamura, Hajime 31, 67, 68, 69, 73, 74
Nakane, Chie 73
Nakayama, Shigeru 22
'negotiated' texts 233
'nested' communities 51
'nested' issues 53, 223, 224
'Nested' structures 50
non-native English speaker(s) (NNS) 38, 40, 42, 45, 46, 47, 51, 53, 54, 60, 76, 78, 79, 97, 125, 127, 222, 224, 226, 227, 233, 235, 243, 248, 250
Oldenburg, Henry 34, 247
Opinion (Op) Step 147, 149, 150, 156, 158, 199, 205
Paltridge, Brian 42, 50, 51
PDisP 147
Phenomenon (Phe) Step(s) 150, 154, 155, 157, 181, 202
philosophies of the East and the West 22
PIntP (Interactive participant) 147
PintP function 149
place discourse community 28
plagiarism 225
power 44
Presenting Steps 139, 143, 144, 151, 152
register 15, 25, 36, 37, 38, 39, 51, 52, 75, 263, 265, 266, 269, 270
research article(s) 34, 35, 43, 75, 77, 82, 84, 247, 248
research science journals 50
review article(s) 36, 38, 41, 42, 51, 62, 81, 82, 88, 89, 90, 91, 92, 93, 94, 95, 98, 99, 103, 109, 115, 120, 121, 125, 161, 237, 243, 246, 249, 250
Review type(s) 207, 221
reviewer(s) 59, 61
reviews 84, 85, 89, 90, 91, 115, 119
rhetoric 19, 23, 64, 71, 248, 249
rhetorical diversity 249
rhetorical organization 102, 221
rhetorical patterns 62, 66
Royal Society 34, 35
RWE (Real World Entity) 146
science 19, 21, 22, 23, 24, 75, 248
science discourse community(ies) 85, 86
science journals 54
scientific facts 244
Sentence Initial Element (SIE) 128, 186
shape knowledge 94
shu-ha-ri 76
SIE (sentence-initial element) 129
 CauseResult (CauseR) 141
 Contrast/concession (Con) 141
 features 137
 Grammatical subject (GS) 138
 Location in space, discourse entity 140
 Question 140
 Time Real world entity (TRWE) 142
social construction of knowledge 20, 249
specialist informant(s) 86, 87, 88, 89, 91, 96, 109, 128, 221, 241, 244, 250
St. John, Maggie Jo 125, 136
status quo
Status Quo reviews 100, 101, 104, 105, 107, 109, 115, 117, 120, 150, 151, 155, 156, 210, 214
Step(s) 125, 127, 128, 136, 150
style manuals 61

substance 41, 43, 47, 48, 74
substance, form and action 42
Sum Step 173, 201
Swales, John 14, 36, 37, 39, 44, 46, 61, 63, 64, 76, 77, 78, 79, 81, 82, 86, 87, 94, 125, 127, 166, 241, 249
Synthesizing Moves 150, 153
systemic functional linguistics 37
Technology (Tech) Step(s) 155, 156, 180, 202
text type(s) 42, 51
textography 36
theory 100, 104, 109
Theory Review(s) 101, 106, 108, 118, 120, 150, 157, 158, 217
tools 224
Topic Step 164, 200
Type I issues 54, 56
Type II issues 56
Type III issues 58
Type III problems 59
vagueness 69, 70, 71
verb(s)
    Past tense 144
    Present tense 143, 180
    verb tense/modality 128, 135, 142
Western science 77
Wordsmith 161
writing support 223, 224, 225
    "write with others' words" 225, 231
    writing-support tool(s) 223, 229, 230, 235

Linguistic Insights

Studies in Language and Communication

This series aims to promote specialist language studies in the fields of linguistic theory and applied linguistics, by publishing volumes that focus on specific aspects of language use in one or several languages and provide valuable insights into language and communication research. A cross-disciplinary approach is favoured and most European languages are accepted.

The series includes two types of books:

- **Monographs** – featuring in-depth studies on special aspects of language theory, language analysis or language teaching.
- **Collected papers** – assembling papers from workshops, conferences or symposia.

Vol. 1 Maurizio Gotti & Marina Dossena (eds)
Modality in Specialized Texts. Selected Papers of the 1st CERLIS Conference. 421 pp. 2001.
ISBN 3-906767-10-8. US-ISBN 0-8204-5340-4

Vol. 2 Giuseppina Cortese & Philip Riley (eds)
Domain-specific English. Textual Practices across Communities and Classrooms.
420 pp. 2002.
ISBN 3-906768-98-8. US-ISBN 0-8204-5884-8

Vol. 3 Maurizio Gotti, Dorothee Heller & Marina Dossena (eds)
Conflict and Negotiation in Specialized Texts. Selected Papers of the 2nd CERLIS Conference. 470 pp. 2002.
ISBN 3-906769-12-7. US-ISBN 0-8204-5887-2

Vol. 4 Maurizio Gotti, Marina Dossena, Richard Dury, Roberta Facchinetti & Maria Lima
Variation in Central Modals. A Repertoire of Forms and Types of Usage in Middle English and Early Modern English.
364 pp. 2002.
ISBN 3-906769-84-4. US-ISBN 0-8204-5898-8

Vol. 5 Stefania Nuccorini (ed.)
Phrases and Phraseology. Data and Descriptions.
187 pp. 2002.
ISBN 3-906770-08-7. US-ISBN 0-8204-5933-X

Vol. 6 Vijay Bhatia, Christopher N. Candlin & Maurizio Gotti (eds)
Legal Discourse in Multilingual and Multicultural Contexts. Arbitration Texts in Europe.
385 pp. 2003.
ISBN 3-906770-85-0. US-ISBN 0-8204-6254-3

Vol. 7 Marina Dossena & Charles Jones (eds)
Insights into Late Modern English.
378 pp. 2003.
ISBN 3-906770-97-4. US-ISBN 0-8204-6258-6

Vol. 8 Maurizio Gotti
Specialized Discourse. Linguistic Features and Changing Conventions.
351 pp. 2003, 2005.
ISBN 3-03910-606-6. US-ISBN 0-8204-7000-7

Vol. 9 Alan Partington, John Morley & Louann Haarman (eds)
Corpora and Discourse.
420 pp. 2004.
ISBN 3-03910-026-2. US-ISBN 0-8204-6262-4

Vol. 10 Martina Möllering
The Acquisition of German Modal Particles. A Corpus-Based Approach.
290 pp. 2004.
ISBN 3-03910-043-2. US-ISBN 0-8204-6273-X

Vol. 11 David Hart (ed.)
English Modality in Context. Diachronic Perspectives.
261 pp. 2003.
ISBN 3-03910-046-7. US-ISBN 0-8204-6852-5

Vol. 12 Wendy Swanson
Modes of Co-reference as an Indicator of Genre.
430 pp. 2003.
ISBN 3-03910-052-1. US-ISBN 0-8204-6855-X

Vol. 13 Gina Poncini
Discursive Strategies in Multicultural Business Meetings.
338 pp. 2004.
ISBN 3-03910-222-2. US-ISBN 0-8204-7003-1

Vol. 14 Christopher N. Candlin & Maurizio Gotti (eds)
Intercultural Aspects of Specialized Communication.
369 pp. 2004.
ISBN 3-03910-352-0. US-ISBN 0-8204-7015-5

Vol. 15 Gabriella Del Lungo Camiciotti & Elena Tognini Bonelli (eds)
Academic Discourse. New Insights into Evaluation.
234 pp. 2004.
ISBN 3-03910-353-9. US-ISBN 0-8204-7016-3

Vol. 16 Marina Dossena & Roger Lass (eds)
Methods and Data in English Historical Dialectology.
405 pp. 2004.
ISBN 3-03910-362-8. US-ISBN 0-8204-7018-X

Vol. 17 Judy Noguchi
The Science Review Article. An Opportune Genre in the Construction of Science.
274 pp. 2006.
ISBN 3-03910-426-8. US-ISBN 0-8204-7034-1

Vol. 18 Giuseppina Cortese & Anna Duszak (eds)
Identity, Community, Discourse.
English in Intercultural Settings.
495 pp. 2005.
ISBN 3-03910-632-5. US-ISBN 0-8204-7163-1

Vol. 19 AnnaTrosborg & Poul Erik Flyvholm Jørgensen (eds)
Business Discourse. Texts and Contexts.
250 pp. 2005.
ISBN 3-03910-606-6. US-ISBN 0-8204-7000-7

Vol. 20 Christopher Williams
Tradition and Change in Legal English. Verbal Constructions in Prescriptive Texts.
216 pp. 2005.
ISBN 3-03910-644-9. US-ISBN 0-8204-7166-6

Vol. 21 Katarzyna Dziubalska-Kołaczyk & Joanna Przedlacka (eds)
English Pronunciation Models: A Changing Scene.
476 pp. 2005.
ISBN 3-03910-662-7. US-ISBN 0-8204-7173-9

Vol. 22 Christián Abello-Contesse, Rubén Chacón-Beltrán, M. Dolores López-Jiménez & M. Mar Torreblanca-López (eds)
Age in L2 Acquisition and Teaching.
214 pp. 2006.
ISBN 3-03910-668-6. US-ISBN 0-8204-7174-7

Vol. 23 Vijay K. Bhatia, Maurizio Gotti, Jan Engberg & Dorothee Heller (eds)
Vagueness in Normative Texts.
474 pp. 2005.
ISBN 3-03910-653-8. US-ISBN 0-8204-7169-0

Vol. 24 Paul Gillaerts & Maurizio Gotti (eds)
Genre Variation in Business Letters.
407 pp. 2005.
ISBN 3-03910-674-0. US-ISBN 0-8204-7552-1

Vol. 25 Ana María Hornero, María José Luzón & Silvia Murillo (eds)
Corpus Linguistics. Applications for the Study of English.
526 pp. 2006.
ISBN 3-03910-675-9 / US-ISBN 0-8204-7554-8

Vol. 26 J. Lachlan Mackenzie & María de los Ángeles Gómez-González (eds)
Studies in Functional Discourse Grammar.
259 pp. 2005.
ISBN 3-03910-696-1 / US-ISBN 0-8204-7558-0

Vol. 27 Debbie Guan Eng Ho
Classroom Talk. Exploring the Sociocultural Structure of Formal ESL Learning.
Forthcoming.
ISBN 3-03910-761-5 / US-ISBN 0-8204-7561-0

Vol. 28 Forthcoming.

Vol. 29 Francesca Bargiela-Chiappini & Maurizio Gotti (eds)
Asian Business Discourse(s).
350 pp. 2005.
ISBN 3-03910-804-2 / US-ISBN 0-8204-7574-2

Vol. 30 Nicholas Brownlees (ed.)
News Discourse in Early Modern Britain.
Selected Papers of CHINED 2004.
300 pp. 2006.
ISBN 3-03910-805-0 / US-ISBN 0-8204-8025-8

Vol. 31 Roberta Facchinetti & Matti Rissanen (eds)
Corpus-based Studies of Diachronic English.
300 pp. 2006.
ISBN 3-03910-851-4 / US-ISBN 0-8204-8040-1

Vol. 32 Marina Dossena & Susan M. Fitzmaurice (eds)
Business and Official Correspondence:
Historical Investigations.
209 pp. 2006.
ISBN 3-03910-880-8 / US-ISBN 0-8204-8352-4

Vol. 33 Forthcoming.

Vol. 34 Forthcoming.

Vol. 35 Vijay K. Bhatia & Maurizio Gotti (eds)
Explorations in Specialized Genres.
316 pp. 2006.
ISBN 3-03910-995-2 / US-ISBN 0-8204-8372-9

Vol. 36 Heribert Picht (ed.)
Modern Applications to Terminological Theories and Applications.
432 pages. 2006.
ISBN 3-03911-156-6 / US-ISBN 0-8204-8380-X

Editorial address:

Prof. Maurizio Gotti Università di Bergamo, Facoltà di Lingue e Letterature Straniere,
Via Salvecchio 19, 24129 Bergamo, Italy
Fax: 0039 035 2052789, E-Mail: m.gotti@unibg.it

www.ingramcontent.com/pod-product-compliance
Ingram Content Group UK Ltd.
Pitfield, Milton Keynes, MK11 3LW, UK
UKHW021852100926
8556IPUK00020B/1180

9 783039 104260